FINDING
AMERICA'S
GREATEST
CHAMPION

Published by Fig Factor Media, LLC.
Printed in the United States of America
Cover Design and Layout by Juan Pablo Ruiz

ISBN: 978-0-9990012-9-5

DEDICATION

KATHY - First and foremost my partner in and love of my life, Thank you for all you that you do for both myself and the sacrifices you have made personally for me, and our children. As we now will soon enter our 4th decade together I cannot imagine life without you.

MOM & DAD - The two people who brought me into this world, and are most responsible for anything that I have done, or succeeded in life I owe to you both. While you left us too soon, Mom your sense of humor and gusto not only for life, but also relationships remain in my core. Dad, your work ethic and attention to not only detail and excellence, but also fairness and honesty have served me well.

MY CHILDREN (Britton, Lindsay & Cameron) - God has blessed me with the most wonderful gift of all, 3 beautiful, honest, independent, hard working adult children. Throughout my life it has been my goal to give you every opportunity, love and support so that you can succeed and prosper. In addition God has blessed me with 6 wonderful grandchildren which are the light of my life, in my senior years.

MY SIBLINGS (Kelly, Amy, Erik & Todd) - You have been so supportive and while we cannot see one another as often as we would like, our connections are forever. Your love and support in all that I do means the world to me.

MY EMPLOYEES - For almost 4 decades I have worked in the family business. We have had the most talented, dedicated and committed employees we could ever hope for. Thank you for being there to make all things possible for not only me, but all of our supportive customers.

MY MENTORS - Throughout my life I have been so fortunate to have been mentored by some of the very best. This set me up in life to feel drawn, obligated and driven to mentor our next generation, whether it be my own children, athletes, or others that I have been introduced to during the journey of life. Thank you for not only the wisdom and insight given to me, but also the concept of paying it forward, to offer sustaining values in a changing world that so desperately needs them.

MY CUSTOMERS - For my entire career I have looked for solutions that would best serve all of my customers. With the trust that we have built with each and everyone of you, I have been able to succeed. Without this success I would not have been able to reach the pinnacle of my craft, and been able to prosper. Thank you to each and every one of you whom put trust into both myself, my company and my employees.

TABLE OF CONTENTS

PREFACE

When I first started in manufacturing, I was totally enamored with recognizing things in everyday life and correlating them to where they were made. Whether it be a water meter on the side of my house, or a pen that I was using, or a snowblower or lawn mower, the list goes on to infinity. We take so many things for granted without even giving it a thought. The United States is one of the largest buying consumer entities in the world. We consume products in massive quantities. Our collective influence is vast.

Early on in my life, buying something made in America was important to me. Whether a car, a cell phone, or something less significant, I always felt supporting someone in the U.S. who made the product should be a priority. To this day, every car that I have owned has been an American car, both personally and for the company.

As you travel through the pages of my book, I ask you to share my passion. Share my curiosity and love for this country in a way that gives back to our American manufacturers and American workers. Our ability to make products ties to our national defense, our security, as well as our standard of living. Making things is in our country's DNA and assures us the quality of life and choices that we have grown so accustomed to. At the same time, there is a sleeping giant in this country, and that is the opportunity that lies ahead for our youth in positions that feature them working with their hands and feeling a tremendous sense of accomplishment, all the while, earning good money. Help me start being a change agent so that as we educate our youth about these jobs... And alleviate the huge student debt crisis that so many of our young people find themselves trapped underneath. There are so many reasons to board this train that will take our country to greatness.

Our culture has become very much blind to the importance of manufacturing in our country. I learned that I had to become more descriptive and elaborate, ie market my profession. So many in our craft continue to do what I did—assume that my peers were not interested or would not find our trade interesting. We can ill afford to do that any longer.

I have titled Chapter Two Manufacturing: America's Gold Unicorn. While you may think the title tells nothing, I think it tells everything. Let me explain. A unicorn is something that very few of us know anything about. It is mythical, thought to never have existed and never much paid attention to. Manufacturing, for many of us, is the same. In the venture capital industry, a unicorn refers to any tech startup company that reaches a $1 billion dollar market value as determined by private or public investment.[1] The term was originally coined by Aileen Lee, founder of Cowboy Ventures. The "gold" element represents the color wealth and that of "a champion." We are in a moment in time that manufacturing can lead us back to prosperity but only if we have the people with the skills to do it. Additionally, there is much to say about the gold standard so many of the contributors in this book have set and the golden rule that we try to live by.

FINDING AMERICA'S GREATEST CHAMPION may be you. It might be the female gender and their place in the workforce, and yes it might even be manufacturing and the good paying careers that they offer. America's Greatest Champion might be the young person whom you have mentored. I feel strongly that more than ever we need to mentor our youth and not just our own children. I have been the benefactor of many of my elders, teachers, coaches and parents' friends guiding me through my life. They stepped in when I needed them to and stepped back when I needed them to allow me to find my way on my own.

I believe in manufacturing. It has been good to my family, and it has been good to this great country. I wrote this book to help the industry that has given my family so much and to help our country recognize and respect the importance of it for our prosperity as a nation. So many company owners and managers are struggling to find skilled workers to grow their businesses and replace baby boomer generation retirees. At the same time, our youth have record levels of student debt and are coming out with degrees that don't have the market value they think they have. Others do not complete their journey through the traditional educational path and see their future to be bleak and in jeopardy. The skilled workforce desperately needs to expand. Salaries are going up as opportunities abound—for those perceptive enough to recognize them. The biggest challenge we have is the perception of these positions. Our culture looks down on them, while our European counterparts hold them in high regard.

I ask you to start a journey with me in this book. Whether you find the Champion in yourself, your child, or a relative, or reach a realization of how important manufacturing is to all of us and our country—*I know your journey will be memorable, enlightening and inspiring.*

— TERRY M. IVERSON

FOREWORD

By Greg Wasson

Besides the fact that our kids are married to each other—my younger daughter and his younger son—and that generationally we're in the same ballpark, it turns out that Terry Iverson and I have a whole lot more in common.

We've both had long-term careers during which we've had ample opportunity to experience innovation and growth in the business world. That said, our "worlds" have been relatively far apart. Terry's been a part of his three-generation-long family-owned, Illinois-based manufacturing/machine tool distribution and rebuilding businesses, which he now heads up. I, on the other hand, fresh out of pharmacy school, spent my next thirty-five years at Walgreens, the nation's largest drugstore chain, starting as a pharmacist and store manager and moving upward through the organization until my retirement as CEO of Walgreens Boots Alliance. Next came my 2016 "relaunch" as founder of Wasson Enterprise, a family office angel-investment company. How much could we possibly have in common from a business perspective? Plenty.

As in-laws, Terry and I, of course, had many occasions to get to know each other through our family connection, but it wasn't until I'd retired from the corporate arena in 2015 and began building my investment business, that Terry and I really got to explore our perspectives on all sorts of topics. We'd get together for breakfast every so often, and that's when I really heard and witnessed Terry's absolute passion for what has become the subject of this book: promoting manufacturing as a career choice and informing parents, educators, and career-seekers (especially high-schoolers and college students) about the mammoth opportunities to be found in the manufacturing field and in entrepreneurship. Having spent his entire working life in manufacturing and building his family business, working through our country's economic, social, educational, and political changes, Terry's had plenty of time to build some pretty intense feelings and thoughts about what's what in the manufacturing field and employment trends in general. As I have during my three and a half decades in the drugstore industry and now in a business that helps start-ups start up, he's observed the role of innovation and disruption in his industry. He's seen, as I have, how creating new products and developing new ways of doing things keeps on changing. He's seen how consumers' needs have changed and that how businesses react to those needs can be the difference between their success and failure. He's seen how the very nature of manufacturing jobs and careers has changed, but not the ways in which people perceive manufacturing as a career path. And on this last point, frankly, he hasn't always been happy with what he's seen and experienced.

Finding America's Greatest Champion: Building Prosperity Through Manufacturing, Mentoring and the Awesome Responsibility of Parenting is truly Terry's manifesto on a subject that's more than just near and dear to his heart. I believe it's his inner core. A family man for whom parenting is part of his essence, he takes that commitment to successfully bringing up young people out into the business and educational world, where he hopes industry leaders and educators will embrace promoting all that manufacturing has to offer. He wants to alter many people's perceptions of manufacturing so they can see its awesome potential and possibilities. And who can benefit? The high-schooler for whom the traditional college degree may not be the best choice. Recent college grads looking for new challenges beyond what they've already studied or been directed into. Those whose creative bent is centered around finding solutions by creating products or looking for ways to innovate how we live.

I absolutely agree. The fact is, we need to steer more people—young people, especially—into manufacturing. That's where the jobs are. That's where the needs are. As Terry writes, the U.S. has more than 12.5 million workers in manufacturing. The industry contributes trillions of dollars to our economy—$2.25 trillion in 2016 alone. Over the next decade, more than 3.5 million manufacturing jobs will likely be needed. And we're talking about everything from assembly line work to cutting-edge technology and robotics to innovative product design and development, and—as the saying goes—much, much more. As our exports increasingly compete on the world stage, manufacturing work that had been underrated as a career choice needs to be brought into the limelight and brought back to the U.S.

To be more competitive, we need skilled and educated people. And so we need to inform not only young folks who are considering the kind of work they're best suited for and can afford to study for, but also their parents and educators down to the high-school level. Terry appeals to parents everywhere to take an active interest in their children's future—to consider the options and alternatives to the traditional four-year college path and opportunities perhaps never imagined. In fact, today, the trades are incredible opportunities that come with less debt—a better return on education, as Terry likes to put it. Think about the things you use in your everyday life. Every one of those items was someone's creation. Someone owned it. Someone manufactured it. That someone was someone's child.

Now, the other aspect of Terry's impassioned treatise that I can really relate to is entrepreneurship. When you dive into the book, you'll see the many people Terry's met and mentored, or admired and approached for their unique perspective on becoming an entrepreneur. And that becoming … that evolution … can be fascinating. Terry's been a keen advocate of entrepreneurship. He's watched his own kids' school mates grow up and enter exciting and diverse fields and then make career decisions that have changed their lives. As someone who's always been an entrepreneur at heart, I get it. I intended to own my own drugstore while I was in pharmacy school, but when the opportunity to take a senior-year internship came along at Walgreens, it ultimately detoured me and became the linchpin into corporate life. But even so, as a drugstore manager who later headed up territories and company divisions until attaining CEO, I had to approach my work as an entrepreneur.

Terry has really put his heart into this book, and it shows. I love his deep allegiance to his family and the business they built long before he was born. He shares his family history throughout the book, telling the story of carrying on the entrepreneurial spirit of Iverson & Company's family founders. When I told Terry that I myself hailed from the greatest entrepreneur and business person I've ever known—my dad, Dick Wasson—he could see that I really understood where he was coming from.

My dad is one of the original I'll-figure-it-out people . . . a real innovator with a pioneer spirit who could make something out of nothing. Following a stint in the army in the early 1950s, he spent years in mechanical repair and servicing before he moved on to selling home improvement products. That's when I learned my first and—ultimately—most valuable lesson in being an entrepreneur. Dad told me, "Find a product or service that people want or need, sell it at a fair price, and service the heck out of 'em!" How true—as I learned repeatedly throughout my career. When Dad decided to buy some open land near our home in rural Indiana in the 1960s and built a family campground (which, although he sold it quite some time ago, still exists today), he made every one of us five Wasson siblings—his campground employees (all of us still kids, I might add!)—adhere to that mantra.

Today, Dad's advice holds true for me as a 100% entrepreneur investing in start-ups through Wasson Enterprise. As any CEO and entrepreneur must, I've always been focused on growth opportunities, driving efficiencies, and cutting costs when necessary. After a couple of true entrepreneurial years, I can honestly say there's a real benefit to being part of a start-up. It's fascinating to experience what it takes to start a company—raising capital, watching every penny, learning quickly the importance of perseverance, and innovation. And to that point, what's needed in manufacturing today is innovation. It's just not the manufacturing of yesterday—and companies need to not only innovate, but innovate the way they innovate.

Terry is a true thought leader, and his book does more than support his point of view, it lays out a strategy for making careers in manufacturing well worth exploring. If you're a parent, consider where your child's special talents might mesh with manufacturing. If you're an educator or a guidance counselor, stay abreast of the opportunities out there for your students post-graduation. If you're in manufacturing, take up the banner for your industry to help it keep going strong.

Our breakfast meetings have proven to be a unique way for us both to leverage our vast experience and talk about how to help spur interest in the opportunities that lie ahead for the next generation in American manufacturing. Companies are built by people and survive by their contributions, creativity, and commitment. Terry's *Finding America's Greatest Champion* is a call to action to young people to take a new or alternative path that may just be the best direction to go.

— GREG WASSON
CEO of Walgreen Co. and Walgreens Boots Alliance, 2009-2015
Cofounder and President, Wasson Enterprise

INTRODUCTION

Writing this book has been an incredible journey. Everything I have done in my life both personally and professionally prepared me to put this piece together. My hope is to make a positive change in our culture for our country. There is a Champion in everyone of our young people, and we need to help each and every one of them discover it.

I have been so very fortunate to know many wonderful friends and acquaintances in my life who have allowed me access to some very talented and insightful opinions and commentary to share in this book. You will find this book very autobiographical in nature. Because I have met so many talented and passionate people, I feel compelled and inspired to have them help tell this story. I conducted over 40 interviews covering the topics of mentorship, parenting, and opportunities and the importance of manufacturing greatness in our country. These are italicized throughout the manuscript. Many of these personalities do not come from the manufacturing sector. Those interviewed come from all sorts of industries, parts of the country and career sets. Professional athletes, media professionals, manufacturing executives, inventors, entrepreneurs, CEO's, and many more. In order to be successful in this book, I feel that I need to reach those not in manufacturing, as well as those who are. In doing so, our culture will hopefully change and, as a result, more young people will be encouraged to enter good-paying and rewarding careers. Many of these are simply not even on young people's radar. They are not informed or even aware they exist.

This book is designed to give you the tools, resources and insight you need to help our next generation become America's Greatest Champion. By reading the table of contents, you will be able to focus on a certain area of the book, a certain personality that you are curious about, or a topic that is more time-appropriate for you than others. The educational sector, job market and parenting models are ever-changing. Perceptions need to be adjusted for what the reality has become. My hope is to bring you on a journey of realization of what your perceptions currently are and what you feel as you travel through the pages of my book. My hope is to enhance your understanding of mentoring, parenting and the importance, significance and opportunities in manufacturing in our lives here as we together are "Finding America's Greatest Champions." Please join with me and become a CHAMPION Now!®.

A LIFE IN MANUFACTURING
The Iverson Path from Norway to U.S. Manufacturing

"The best lesson my dad taught me was always be totally honest with your customers and totally honest with your employees. Don't fib, don't stretch the truth, be fair, be equitable, and be honest."

– Jerry R. Iverson, Chairman Iverson & Company

This country has been formed with many immigrants from countries near and far. Europe and Scandinavian countries have had a long history of craftsmen, artisans and fisherman. My family on my father's side originated from Norway. My great-grandfather came over on a ship and never got back on when it returned home. Like many immigrants coming to the United States, he started a family—hoping for a better life. The manufacturing sector has reaped the benefits of many talented and skilled tradesmen, coming over from lands far away. Our forefathers lived life with a code of honesty that guided them in a way that led to success and happiness.

Think about it. Where did the people who came before you live and work? What were their ambitions? What did they sacrifice to make a better life for you? This is part of your story.

That's why I must begin here, with the legacy that manufacturing has left for me.

My family has been in manufacturing for over ninety-three years. My grandfather, Edward Iverson, worked for a then small machine tool builder in Chicago back in 1925. Keep in mind that this was shortly before the Great Depression (1929-1932). He had to quit school at sixteen years of age, when his father died at a young age, to work full time to support his five siblings and mother. He started at this company by sweeping the floors. I know this "humble beginnings" story may seem cliché, but it is true. His supervisor soon realized that he possessed talents, not only in math and science, but also drafting. Grandpa Senior (Iverson) quickly learned to be vital in the company and rose in importance in his department. In fact, when the company was bought out in 1931, there were only three people who were offered a position in the plans to move east to New York State. He was, in fact, the only one to decline and, as a result, stayed back to represent the manufacturer. I have thought many times about this bit of family history. Here my grandfather was, in the middle of the Depression (1929-1932) in Chicago, and the company he worked for was purchased. He had to make a decision to move to New York or stay. He made the tough decision to stay and not uproot the entire family. This bold move and him being the major breadwinner at such a young age were the first indicators I can remember of how important a sense of family is and continues to be in the Iverson namesake.

When this machine tool builder moved to upstate New York, my grandfather stayed behind and became their representative for the Midwest. (You can say we were their first distributor in the world.) He was quite good at what he did and grew to know many industry leaders in the manufacturing sector. My father, Jerry, started working for him in 1958 and Uncle John from 1965-1972 in the capacity of machine tool sales. My Uncle Ed ended up starting his own subcontract machine shop called Chucking Machine Products – machining parts for Bell & Howell. They have now celebrated their 60th year in business and now specialize in high precision aerospace, defense and military subcontract machine work. John after some years with much success in the Wisconsin market, bought a small shop called Custom Products. We will talk about both of them later. My father Jerry stayed with the company and perfected his crafts of machine tool sales and rebuilding. He has been my greatest mentor to this day, and I thank him for teaching me so much about life and work. My father is very detail-oriented and financially astute. The way his mind works is very analytical and logical. His greatest words of advice were:

"Terry, you have everything if you have family and your health. Everything else is secondary."

For years I didn't understand the significance of Dad's advice until Kathy and I came to grips with a serious surgery for a newborn child. I will never forget my father's advice and the feeling of helplessness while our child went in for open heart surgery. Appreciating what we have, not taking it for granted, and finding the inner Champion in everyone inspires me to write this book.

When my grandfather started our machine tool distributorship, he traveled to New York State and purchased a small indicator company by the name of Geneva Gage. He and his partner, Pete Flauter, moved the company to Illinois and changed the name to Chicago Dial Indicator. They decided early on that private label manufacturing of dial indicators would be a big part of their business model. They also continued making products under the CDI name as well. All three brothers worked in varying degrees of involvement and contributions in Chicago Dial Indicator (CDI). By 2004, Jerry, John and Ed had all relinquished their ownership. Today, my brother Erik runs and owns the company.

MY JOURNEY INTO THE FAMILY BUSINESS

I began the third generation in our business when I started full time in 1980. During the summers of the early '70s, around the start of my high school career, I started packing collets, cleaning machines, taking apart spindle motors, cleaning the shop and mowing the lawn. We were in Chicago, where we had been for several decades. When we painted machines, the smell of the fresh Sherwin Williams 7B paint would permeate through the shop. To this day, when we paint machines, the smell brings back images of when I was fourteen and starting in the family business.

After I got married in 1980, I entered into the sales side of the business in Wisconsin. I would leave the house every day bright and early to make sales calls. Each machine shop I entered was either a screw machine shop with a very strong cutting oil "aroma," or a CNC shop (computer numerical control) with a differing water soluble coolant smell. At the end of the day, I made the drive home. Upon arrival, I learned that my suit would absorb these industrial olfactory identifiers, not necessarily appreciated by my better half. "You smell like a machine shop." Yes I do. To me, it was the smell of money, opportunity—and the pride of helping customers make something that was useful to the world at large. Today's shops feature a great deal more mist collection equipment and are cleaner. They no longer have the strong oil smell throughout. I found it fascinating seeing locks, water meters, motorcycles, guidance systems and much much more being made from the ground up.

My (much) younger brother Erik also worked for us from 1995 to 2002. Erik left our company to go to Chicago Dial (mentioned above) where he now runs (and owns) the company as president. Everyone in our family is proud of the Made in America moniker and passion. We have believed in the ingenious and productive means that we have here in the U.S. Chicago Dial is a good example of a small manufacturer competing against some of the giants—in this case in the quality control and inspection world.

FOURTH GENERATION – TRYING THE FAMILY BUSINESS

We have had the fourth Iverson generation represented by our oldest son Britton (2004 to 2007), and our daughter Lindsay Weglarz (2008-2013) when they worked for the family business. Our youngest, Cameron, spent a few summers working for us before he decided not to join the family business. After he married, he was able to be mentored by his father-in-law Greg Wasson, a very successful executive (retired CEO of Walgreens – 18th Fortune 500 company) and author of the forward for this book. I think the time that Brit spent in our business gave him the tools to do very well in the banking (21st Fortune 500 company) industry. He has found his niche here and is doing very well. Our daughter Lindsay surprised herself in that she actually had found it to be very interesting, before excelling in the marketing department of an insurance company (84th Fortune 500 company). I am so very proud of all three of our children and honored to mentor each of them as their dad.

For years I had a tough time as the only sibling "going the distance" in the family business. There were ample chances to move on to other opportunities. Not being one to not finish a job, and being loyal to the family, I stayed in the positions earned through the years. Then in 1999, came a week at the Harvard Business School. It was a strategic finance week-long seminar. I found that people were interested in what I had to say, from the perspective of running our small family machine tool business. The thing that impacted me the most was a young lady from South America. I mentioned being the only child in the family business. She commented, "Oh, so YOU are the chosen one!" What?" I thought. "In our country, it is an honor to be the chosen one — to run the family business." she said. I thought long and hard after hearing her commentary. That one encounter allowed me to continue down the path of what now is my 38th year. I always felt a strong obligation to my family and our origins. This was a third party confirmation that, in fact, it was an honor, and I should see it as such.

My wife Kathy works with me. I am very proud of her and have all the confidence in the fact that she can accomplish great things in anything she does. I also feel that way about ALL three of our children. I am a big advocate for women in our industry, as we will later talk about in Chapter 11 in detail.

"ED'S BROTHER JOHN" AND "JOHN'S BROTHER ED"

My two uncles went on to start their own subcontract machine shops. Uncle Ed started Chucking Machine Products. Ed probably has the most clear resemblance physically and personality-wise to my grandfather. He has done well with his ability to process parts and run a world class subcontract machine shop. I would say that all three of the brothers and myself have possessed/developed that skill set. I find it hilarious that my dad always refers to his brother Ed as John's brother and John as Ed's brother. I think this alludes to the fact that sometimes brothers can be a little too close, and sibling rivalry comes into play. Of course Ed and John do the same to Dad. All of this is packaged with both love and respect amongst the three.

Uncle John bought Custom Products in 1972. Both uncles have been enormously successful in machining parts for some of the world's largest and most easily recognized manufacturers. I worked for John in the summer during college soon after getting married. My father asked me to join the family business, after about six months working for my uncle. John was probably the greatest salesman in the family. He was known for a lot of things—a very generous person, but he knew when and how to take risks in business. Eventually, I learned to take risks from his example, although I will never be the big risk taker that he ultimately became. After building his company of over 700 employees, he ended up selling it to a large casting house with many locations nationally.

As I outlined in the above paragraphs, my family has been intimately involved in the "Art of Manufacturing." Both Iverson & Ternstrom (Company) as well as Chicago Dial Indicator have provided careers and a great deal of sustenance (and in some cases wealth) for all families involved. My dad's Uncle Stanley Ternstrom (his mom's brother) started the rebuilding side of the business back in the 1940's. Many of the Ternstrom family has, and continues to, work for us.

Although most of us have very moderate taste (Uncle John had a taste for the good life*), we have been able to provide a good and comfortable way of life for our families. All the while, for most of our careers, working in technologically advanced shops that were not only safe but also clean.

John was known for being a very social person. His parties at his home, were known to be the talk of the town with upwards of 300+ people. Going to these parties was quite a spectacle. John would call in local youth orchestras to play while adorned in formal attire for entertainment for his guests.

I have had the extreme honor to meet so many good, talented, honest and hard-working people. I am fascinated with meeting people. I always have been. (My mom was the same way.) Maybe because we moved sixteen times in my youth. I was ALWAYS meeting new neighbors. I had to sharpen my skills to become a "social junkie." In this book, you will meet some fascinating people— some in manufacturing but many of whom are not involved in manufacturing. I feel that it is relationships that make the world go round. Even in this time of technology, we all yearn to meet and interact with other people. I have had the best job in the world because every day I get to meet new people and get my fix of interaction. It also allows me to use my engineering skill set to help them make their products better, faster and more accurately.

CHAMPION NOW!® A VISION – A MOVEMENT IS BORN!

I began coaching and mentoring young people when our children were young. I eventually ended up retiring from coaching when my youngest son,

Cameron turned eighteen, and several of his teammates went on to play Division III soccer. Coaching had run its quarter century course, yet still I felt compelled to mentor youth. I was so fortunate to have mentors when I was young, and I felt a great obligation to pay it forward.

Around 2009, Bryan Albrecht (President of Gateway Technical College) invited me to join the CTE (Careers in Technical Education) Foundation in Washington DC. Through this involvement, I was able to attend and make presentations at annual conferences in both Charlotte and Nashville on how to have model partnerships with industry and technical colleges pertaining to manufacturing advances. I also got involved in a Leadership Forum in Washington at Union Station. On my way to one of these meetings in Washington, I myself became inspired.

I thought, "How do I coin a slogan or an organization to change people's perceptions?" I already had pitched a television idea, but that alone did not seem to be enough. There had to be a message alongside anything that talked about manufacturing. I struggled with the terms Changing and Perceptions and Manufacturing. After all, that is the dilemma facing all of us trying to attract "new blood." While scribbling on a napkin on a plane, I toyed with the words CHANGE – MANUFACTURING – PERCEPTIONS. Somehow I came up with C-H-M-P. I then found my way to the word CHAMPION – Change How American Manufacturing's Perceived In Our Nation. I was stunned with the sublime positive implications with the word alone.

I added the Now to it to initiate action to move—immediately. In other words, there is a crisis in our country and that is that no one thinks, is exposed to, or is allowed to pursue careers in manufacturing. As a result, our workers, managers and owners are getting older. There is no succession plan. Nepotism is the only plan there is. That, by itself, is not really a sustainable one.

The manufacturing and machine tool industry have an annual association meeting that is collectively held for four associations. This first collective meeting was named MFG or Manufacturing For Growth. It was held in Phoenix, and I decided that we needed to promote and introduce CHAMPION Now!® to the machine tool industry. Pete Borden, (then President of the AMTDA American Machine Tool Distributors Association) decided to allow me to write and also print a two-page call to action letter to the industry. I printed 700 copies, one for each person in the general session. That was very cool and unbelievably supportive for such a naïve and not very well-known notion of CHAMPION Now!® initiative. What I didn't know was that Pete had decided to give me the first Manufacturing Initiative Award at this 2010 association meeting. I was very surprised and honored. I never thought that something like this would get anyone's attention, especially from our association president. The fact that he felt strongly that this was a good thing made me feel fantastic and not want to give up on the concept of being a unifier in the drive to encourage everyone to know more about manufacturing careers.

CHAMPION NOW!® – SIGNING DAY AT TECHNICAL SCHOOLS

What if we celebrated our youth's decisions to go into manufacturing with a celebration and signing day with notable media recognition? We should celebrate them being Manufacturing Champions!

Recently in Henrico County Public Schools in Virginia, Director of Career and Technical Education, Mac Beaton, decided to celebrate their technical education graduates by signing letters of intent for employment. This is similar to when high school athletes are celebrated on National Signing Day when they commit to go and participate in collegiate athletics for their chosen college.1 What a great concept that should be replicated all across the nation. I envision a CHAMPION Now!® Days of Champions event.

One obstacle is the radio silence we often experience about manufacturing. Or I should say what we don't see or hear about manufacturing! It's the best kept secret that young people oftentimes are simply not aware of. Yet, it IS one of America's greatest opportunities—one of her Greatest Champions. Next up, we'll take a surprising look at the facts, highlights and myths of manufacturing, why it is so crucial to our country's future and our culture's profound—and surprising—influence on manufacturing.

MANUFACTURING AMERICA'S GOLD UNICORN

". . . so when I hear people go, 'You guys make your product outside the U.S.' Really? You buy products (made) outside the U.S. Stop buying it outside the U.S. And we can start making it in the U.S."

– Tony Schumacher, NHRA Top Fuel Champion

The U.S. manufactures just under 20% of the world's goods. Yet, our culture often gives the impression we aren't a manufacturing populace and that manufacturing is not important. Time to WAKE UP. Our culture needs to be engaged in WHERE things are made, HOW things are made and WHY it is important for us to care!

The point is that manufacturing creates many things. The obvious interpretation is in its definition. According to Merriam-Webster it is as follows.[1]

DEFINITION OF MANUFACTURE

1: Something made from raw materials by hand or by machinery

2: The process of making wares by hand or by machinery especially when carried on systematically with division of labor. b: a productive industry using mechanical power and machinery

3: The act or process of producing something

There is a less obvious meaning behind the phrase "manufacturing creates." This is a trademarked term that I came up with to tell this story. People's interpretation may be very naïve, as in, of course, manufacturing creates . . . isn't that the definition of manufacturing itself? Isn't that obvious? But my explanation goes further than the products that are the lifeblood of any manufacturer.

Manufacturing create$™ Opportunities

Manufacturing create$™ Jobs

Manufacturing create$ ™ Careers

Manufacturing create$™ Wealth

Manufacturing create$™ Stability

In this country, we have gotten away from this belief, or the basic understanding, of what has made this country great. Much of the media like to say that we have become a service based country. I say that is not only not true, but believing that to be the case, can become a self-perpetuating negative prophecy.

The USA manufacturing economic engine is the eighth largest economy in the world. We still manufacture approximately one fifth of the world's goods. Does that sound like manufacturing is dead? In this country, the biggest challenge manufacturers have to help them grow is to have an adequate supply of talent to make it happen. This is one of the great opportunities for our young people in this country.

STEPPING UP FOR MANUFACTURING; THE GAUNTLET THROWN DOWN

In March 2009, I was invited to talk for five minutes on the "Monster and Money" television show in Chicago. At the time, a friend from high school (Don Dupree—we will talk to/about him later) was the producer of this program and knew of my manufacturing expertise and asked me on the show. It was a live broadcast on Channel 2 Chicago local TV—at a very early time of the morning. I think I was on somewhere around 5:40 a.m. I talked about my CHAMPION Now!® organization that I founded to promote manufacturing careers and that manufacturing was still in need of young talent. They give you five minutes to talk and ask you questions on live television. Of course I was nervous, and a bit half asleep as well. I doubt I had adequate coffee intake that morning. There were four hosts on the show: ex-Chicago Bear and sportscaster Dan Jiggetts, radio personality Mike North, Mike Hegedus of Hegedus World, and financial expert Terry Savage. (Dan's daughter and my oldest son went to high school together. She is a very successful news anchor in Chicago now).

I wasn't cued into what questions were going to be asked or how the interview was going to go. I mentioned my organization and how I had become a self-proclaimed advocate for manufacturing jobs. Mike North talked about the fact that a lot of the manufacturing companies had moved to the suburbs. Dan Jiggetts commented on the fact that, growing up, he heard that manufacturing jobs were good paying jobs—years ago. Then came the comment I never saw coming from Terry Savage ... when she said:

" I find it very counterintuitive that you are sitting here saying that today there are jobs available and that we should tell young people to pursue careers in manufacturing!"

Needless to say, I was floored. Talk about being put on the spot! The response I gave was adequate but certainly not convincing to the degree that I am capable of. The answer I gave was simply, "Terry, there are manufacturing jobs that cannot leave American: soil—defense and medical to name a few. We will always have to make the most complex, advanced and most proprietary products in our country." The component of the response that I left out was how the aging population in manufacturing exceeds the number of products and jobs leaving for China. Although it is true that a great deal of low-cost products and processes are going to China, but with automation and advanced production techniques and equipment, we can compete globally. The average age in manufacturing is fifty plus years of age. With the latest recession, many of the baby boomers working in manufacturing have postponed their retirement, but the catastrophic flow of retirees out of manufacturing is inevitable. If the resultant massive need for workers is not already here—it is certainly soon to come!

It was probably on this show, when Terry asked me this question, that I subconsciously decided I had to write my book.

A CONVERSATION WITH DON DUPREE
EMMY WINNING PRODUCER - SWIMMING UPSTREAM AGAINST THE MEDIA

Back in high school at Bolles in Jacksonville, Florida, I played several sports, one of which was football. When I made the varsity football team as a junior, one of my teammates was a lightning-fast running back by the name of Donnie Dupree. He was an awesome athlete with a GREAT sense of humor. At one point, he worked for Oprah Winfrey then went on to become the producer for the Regis and Kathie Lee and Siskel and Ebert shows. As mentioned above, he had a stint at Channel 2 here in Chicago until he finally got a big chance with his own show −"A Piece of the Game." This is a show currently running that focuses on sports memorabilia and the stories behind the item with the owner. They have a tagline "WhaddaYa Got?" Don has been awarded four Emmys for his hard work and commitment to the industry. He certainly deserves it. His story and success are a true testimony to his character, much like others I have chosen to be interviewed and included in this book. Don talks about the culture and the role the media plays in manufacturing.

One of the biggest obstacles you face is what's in the media every day. It's played out as a storyline that manufacturing jobs are dead. They're going overseas. Factories are closing. People are losing their jobs. Small towns are drying up. It's the wrong place to be.

Those misconceptions are so deep seeded. It's hard to convince anybody to do anything. It might be an opportunity. I think you're swimming upstream against the media that paints the picture every day. It's jobs that are available. Jobs in manufacturing that are interesting. (They) are not what people think they are, you know, grease and grime.

Some of these jobs involving manufacturing and technology are more advanced and require training. I think there's a huge opportunity there. I think unless something's done, people are going to begin looking at manufacturing not as their friend, but the enemy. They'll look at automation as taking away jobs. I think it's one of the great things facing our country over the next generation. Are we going to be ahead of that curve or behind it? I think training people to do automation, having it work for us, and creating jobs from it is critical on so many levels.

You've got jobs that are well-paying that are going unfilled. People don't know about it. You've got opportunities to train people to take these jobs. I think that is about as powerful as you can get.

You've got something much more powerful. You're selling jobs, employment, and life-changing events. You're coming at it from a position of

35

strength. Most people, like me, are going to be surprised to find out training, jobs and all the things you have to offer are available The information about U.S. manufacturing being number eight (economy in the world) shocks me. I bet most people would be. All you hear on TV is the other side of things.

I really like you leading this movement, the manufacturing movement. I think what we're talking about is reaching people on social media and doing things like that, it's even bigger.

A CONVERSATION WITH WARREN YOUNG
CEO of ACME INDUSTRIES
ENTREPRENEURSHIP OPPORTUNITIES ABOUND

With my relationships with schools, I get asked to speak fairly often. Another industry member, Warren Young of Acme Industries, was also asked to speak at Harper College, which is a well established community college in the northern suburbs of Chicago. (Harper is situated on 200 acres with 56,000 students.) More than once, Warren and I ended up on the same panel, sitting next to one another. In listening to his story, I was inspired. Here is a guy who was pushed aside by a manufacturing company. His gut tells him not to give up and double down on his career. He goes out and acquires a company and becomes the boss. He makes the decisions and controls his own destiny. That is so cool. This exemplifies the endless entrepreneurship opportunities that manufacturing has to offer.

I have been in manufacturing my whole life. I realized, when I graduated with my engineering degree, that I was quite different than a lot of other engineers, who maybe thrived at the desk, doing their calculations. Maybe it's part of my upbringing from the farm. I liked to be out touching things, doing things, seeing results. It just looked to me like manufacturing could be full of that. It could entail solving problems that weren't necessarily strictly analytical.

My youngest son decided when he was picking his major, he was going to take industrial management and that one of his minors was going to be manufacturing management. My wife says, "Why would you want to do that? You know, manufacturing isn't very viable anymore."

That really struck home. How could someone in my same household say that to her offspring after her husband was still making his career in manufacturing? We all know there is truth to the sense that a lot of manufacturing has left this country. Just like a lot of people have come off the farm. There is an analogy there. That doesn't mean that one should take history, or political science, and just hang out on the fringe of the employed world. Not only is manufacturing NOT dead, it is being revitalized. There is a bit of a manufacturing renaissance! I think that we're going to see in the next few years, based on the tax reform that's happened, there's going to be a lot of vibrancy in manufacturing. Our biggest challenge is going to be the task of marketing manufacturing as a fulfilling and well-paid career opportunity. If we can't find the people who are willing to come into manufacturing because they think there is no opportunity there, we are going to be dead.

In 1991, I was transferred to Chicago by the company I worked for at the time. After six years with three different companies, I was thrown onto the beach, as some would say. Instead of looking for another job, I took the advice of one of the guys I had worked for in a prior life. He said, "Why don't you find a company to acquire? I did it, and you can too!" At the age of fifty, not wanting to move my family again and being at a level that while I wasn't king of the world, it was high enough that the jobs at that age for, what I wanted to do, weren't that numerous. It's almost by default that I set out on that journey: okay, I'm gonna go do this.

With the help of an intermediary, we started looking for a certain size of company with a geographic proximity that wouldn't require me to move my family. I was also looking for the type of business that would match up with my experience and my background, which had been spent in the manufacturing of precision industrial equipment. In the search for a company that could be the next thing to go to, I came across Acme Industries about two months into my mission.

Fortunately, this particular company matched up with my background. While I had never been responsible for selling machined parts before, the machining of precision parts was always part of the manufacturing of the equipment that had been part of my responsibility, making printing equipment or packaging equipment or mechanical seals, or whatever.

Acme had seventy-five people at the time, doing $18 million in sales. They were located in 40,000 square feet of space in two building in Des Plaines. They had just put in the last piece of equipment that would fit in the buildings, and there was no room to do anything to grow the business. At our high-water mark, we have had up to 250 people. Today, we have about 180 people working in two buildings that have a total of 270,000 square feet of space in Elk Grove Village, IL. For the year just completed, we did around $50 million in sales, and we think next year we will do about $70 million.

I think the analogy to agriculture is still appropriate. There have been studies done from 1900-2000. The fact that there are less manufacturing employees than there used to be is due to the productivity improvements, innovation, automation, and technology that have impacted manufacturing. It's also due to the offshoring of some amount of work. I think if you want to look at the glass half full, we can't and shouldn't begrudge some of the manufacturing that's left the country. That wasn't where our sweet spot was for our country's manufacturing base. I think that what we need to emphasize is the fact that there is still a lot of good manufacturing here in the U.S., and we can, and should, continue to build on it.

If we can't do those higher end jobs that have the complexity, the need for technology, the need for innovation and entrepreneurship—if we can't win that game—then shame on us.

Below you'll see some of the fascinating and little known facts about manufacturing that will hopefully open your eyes. These twenty facts should bring to light so many facets that most people do not know. The National Association for Manufacturing has provided this information that is so enlightening and crucial for everyone to understand the reality of manufacturing in the United States.[2]

"TOP 20 FACTS ABOUT MANUFACTURING"

1) In the most recent data, manufacturers contributed $2.25 trillion to the U.S. economy in 2016. This figure has risen since the second quarter of 2009, when manufacturers contributed $1.70 trillion. Over that same timeframe, value-added output from durable goods manufacturing grew from $0.87 trillion to $1.20 trillion, with nondurable goods output up from $0.85 trillion to $1.00 trillion. In 2016, manufacturing accounted for 11.7% of GDP in the economy.[3]

2) For every $1.00 spent in manufacturing, another $1.89 is added to the economy. That is the highest multiplier effect of any economic sector. In addition, for every one worker in manufacturing, there are another four employees hired elsewhere.   With that said, there is new research suggesting that manufacturing's impacts on the economy are even larger than that if we take into consideration the entire manufacturing value chain plus manufacturing for other industries' supply chains. That approach estimates that manufacturing could account for one-third of GDP and employment. Along those lines, it is also estimated that the total multiplier effect for manufacturing is $3.60 for every $1.00 of value-added output, with one manufacturing employee generating another 3.4 workers elsewhere.[4]

3) The vast majority of manufacturing firms in the United States are quite small.In 2015, there were 251,774 firms in the manufacturing sector, with all but 3,813 firms considered to be small (i.e., having fewer than 500 employees). In fact, three-quarters of these firms have fewer than twenty employees.[5]

4) Almost two-thirds of manufacturers are organized as pass-through entities. Looking just at manufacturing corporations and partnerships in the most recent data, 65.6% are either S corporations or partnerships. The remainder are C corporations. Note that this does not include sole proprietorships. If they were included, the percentage of pass-through entities rises to 83.4%.[6]

5) There are nearly 12.5 million manufacturing workers in the United States, accounting for 8.5% of the workforce. Since the end of the Great Recession, manufacturers have hired more than one million workers. There are 7.8 million and 4.7 million workers in durable and nondurable goods manufacturing, respectively.[7]

6) In 2016, the average manufacturing worker in the United States earned $82,023 annually, including pay and benefits. The average worker in all nonfarm industries earned $64,609. Looking specifically at wages, the average manufacturing worker earned more than $26.50 per hour, according to the latest figures, not including benefits.[8]

7) Manufacturers have one of the highest percentages of workers who are eligible for health benefits provided by their employer. Indeed, 92% of manufacturing employees were eligible for health insurance benefits in 2015, according to the Kaiser Family Foundation. This is significantly higher than the 79% average for all firms. Of those who are eligible, 84% actually participate in their employer's plans, ie, the take-up rate. Three are only two other sectors— government (91 %) and trade, communications and utilities (85%) that have higher take-up rates.[9]

8) Manufacturers have experienced tremendous growth over the past couple decades, making them more "lean" and helping them become more competitive globally. Output per hour for all workers in the manufacturing sector has increased by more than 2.5 times since 1987. In contrast, productivity is roughly 1.7 times greater for all nonfarm businesses. Note that durable goods manufacturers have seen even greater growth, almost tripling its labor productivity over that time frame.

To help illustrate the impact to the bottom line of this growth, unit labor costs in the manufacturing sector have fallen 8.4% since the end of the Great Recession, with even larger declines for durable goods firms.[10]

9) Over the next decade, nearly 3.5 million manufacturing jobs will likely be needed, and 2 million are expected to go unfilled due to the skills gap. Moreover, according to a recent report, 80% of manufacturers report a moderate or serious shortage of qualified applicants for skilled and highly-skilled production positions.[11]

10) Exports support higher-paying jobs for an increasingly educated and diverse workforce. Jobs supported by exports pay, on average, 18% more than other jobs. Employees in the "most trade-intensive industries" earn an average compensation of nearly $94,000, or more than 56% more than those in manufacturing companies that were less engaged in trade.[12]

11) Over the past twenty-five years, U.S.-manufactured goods exports have quadrupled. In 1990, for example, U.S. manufacturers exported $329.5 billion in goods. By 2000, that number had more than doubled to $708.0 billion. In 2014, it reached an all-time high, for the fifth consecutive year, of $1.403 trillion, despite slowing global growth. With that said, a number of economic headwinds have dampened export demand since then, with U.S.-manufactured goods exports down 6.1 % in 2015 to $1.317 trillion.[13]

12) Manufactured goods exports have grown substantially to our largest trading partners since 1990, including to Canada, Mexico and even China. Moreover, free trade agreements are an important tool for opening new markets. The United States enjoyed a $12.7 billion manufacturing trade surplus with its trade agreement partners in 2015, compared with a $639.6 billion deficit with other countries.[14]

13) Nearly half of all manufactured goods exports went to nations that the U.S. has free trade agreements (FTAs) with. In 2015, manufacturers in the U.S. exported $634.6 billion in goods to FTA countries, or 48.2% of the total.[15]

14) World trade in manufactured goods has more than doubled between 2000 and 2014—from $4.8 trillion to $12.2 trillion. World trade in manufactured goods greatly exceeds that of the U.S. market for those same goods. U.S. consumption of manufactured goods (domestic shipments and imports) equaled $4.1 trillion in 2014, equaling about 34% of global trade in manufactured goods.[16]

15) Taken alone, manufacturing in the United States would be the ninth-largest economy in the world. With $2.1 trillion in value added from manufacturing in 2014, only eight other nations (including the U.S.) would rank higher in terms of their gross domestic product.[17]

16) Foreign direct investment in manufacturing exceeded $1.5 trillion for the first time ever in 2016. Across the past decade, foreign direct investment has jumped from $569.3 billion in 2006 to $1,532.4 billion in 2016. Moreover, that figure is likely to continue growing, especially when we consider the number of announced ventures that have yet to come online.[18]

17) U.S. affiliates of foreign multinational enterprises employ more than 2 million manufacturing workers in the United States, or almost one-sixth of total employment in the sector. In 2012, the most recent year with data, manufacturing sectors with the largest employment from foreign multinationals included motor vehicles and parts (322,600), chemicals (319,700), machinery (222,200), food (216,200), primary and fabricated metal products (176,800), computer and electronic products (154,300) and plastics and rubber products (151,200). Given the increases in FDI seen since 2012 (see #15), these figures are likely to be higher now.[19]

18) Manufacturers in the United States perform more than three-quarters of all private-sector research and development (R&D) in the nation, driving more innovation than any other sector. R&D in the manufacturing sector has risen from $126.2 billion in 2000 to $229.9 billion in 2014. In the most recent data, pharmaceuticals accounted for nearly one-third of all manufacturing R&D, spending $74.9 billion in 2014. Aerospace, chemicals, computers, electronics and motor vehicles and parts were also significant contributors to R&D spending in that year.[20]

19) Manufacturers consume more than 30% of the nation's energy consumption. Industrial users consumed 31.5 quadrillion Btu of energy in 2014, or 32% of the total.[21]

20) The cost of federal regulations fall disproportionately on manufacturers, particularly those that are smaller. Manufacturers pay $19,564 per employee on average to comply with federal regulations, or nearly double the $9,991 per employee costs borne by all firms as a whole. In addition, small manufacturers with less than 50 employees spend 2.5 times the amount of large manufacturers. Environmental regulations account for 90% of the difference in compliance costs between manufacturers and the average firm.[22]

This website is every eye-opening to anyone not familiar with manufacturing and also to some of us who are. These statistics allow people to look at manufacturing as an opportunity for those looking for a pathway into something that maybe they never knew existed.

MANUFACTURING'S ENDEMIC PROBLEMS

One of the biggest unknowns in this country is that there are 2.7 million baby boomer workers in the United States manufacturing workforce set to retire between 2015-2025.[23] The average age of a manufacturing employee is in their fifties. According to the Manufacturing Institute, over the next decade, nearly 3.5 million manufacturing jobs likely need to be filled. The skills gap is expected to result in 2 million of those jobs going unfilled.[24] This is at a time that unemployment has hovered around 4%.[25] At the same time, the media preaches NOT to go into manufacturing related fields. Those of us in the industry have an obligation to correct the course that seems to be set. In order to be able to allow for manufacturing to continue to drive the financial stability of our country and all those employed by it, we need to encourage the brightest, most energetic and passionate to go into the manufacturing field. Boeing is very blatant about how many engineers that they need to be able to produce their planes. Many companies have no idea how they are going to staff their production needs in the upcoming years. One ThomasNet article claims that Boeing has a backlog of 5,864 airplanes; that is $135 billion and works out to about seven years of production work.

In the week before the Bourger International Air Show in Paris, Boeing revised its Current Market Outlook 2013-2032, predicting a doubling of the international commercial fleet from the current level of 20,310 today to 41,240 by 2032. Removing retiring vehicles, Boeing predicted that 35,000 of these airplanes would be new, a value of $4.8 trillion. Considering the growing demand for fuel efficiency and the rising price of fuel, 24,670 of these airplanes are expected to be single-aisle aircraft seating between 90 and 230 passengers, while just 760 will be wide-body jets for more than 400 passengers.[26]

It is no secret that many American manufacturers got complacent and fell asleep at the switch. My dad would tell me deliveries in the machine tool business got to the point where it took twelve to twenty-four months to get a very basic standard machine tool in the 1960s and 1970s. This opened the door to aggressive machine tool builders from Japan and Taiwan to have more competitively priced products, as well as in stock deliveries. Machine tool builders in these countries saw an awesome opportunity. They seized the moment. As a result, the machine tool business was never the same in the U.S. Many would say that it is now to a point of extreme risk to our national defense. What would happen if we went to war with some of the countries that we now consider our allies? Many of these could be the same countries that supply machine tools to make the defense mechanisms. Now we could possibly find that we would be unable to keep these machine tools operational in order to supply our arsenal of defense, to our military. A strong machine tool foundation in our country is vital.

The automotive industry also got apathetic from the standpoint of quality. We made inferior products and just expected the U.S. market to not only accept it, but also pay more for it. We did not listen to the consumer as to what they wanted, expected and needed. The Japanese came in with reasonably priced, economically-run automobiles that featured a great deal more miles to the gallon. The American automotive executive allowed for products to be designed, built and manufactured that were primarily status-defining, luxury cars that cost a great deal of money and were very inefficient. Now we have executives, like Mary Barra at GM, who think differently. Not only has she shown that the glass ceiling can be shattered for her gender, but she also will lead the automotive giant into new more responsible and responsive designs to build the cars for the future. The point is unfortunately that we got what we deserved. We didn't listen to the consumers in both cases. What would be worse yet is if we didn't learn from our mistakes as a culture and instead became more competitive and more responsive to the market demands. We also need to be more responsible in attracting, creating and developing the next generation of manufacturing workers, employees and leaders.

AUTOMATION: ENEMY OR ADVOCATE

There is heated debate about the balance between the implementation of automation in order to compete globally, and displacing human labor as a result which would seem to be counterproductive. I contend that we MUST automate and utilize the highest level of technology in our manufacturing processes. This is the world that I have resided in for almost forty years. This will, in fact, replace workers who do very rote, mundane, mechanical tasks. These are also the same tasks that are so easily replicated in China and other far east countries because of their exceedingly low labor rates. Both China and India, with over a billion people in each country, constitute a surplus of labor. As Germany and Switzerland have done for decades, we need a higher level of skill for the workers required to program, implement, service, maintain and improve our highly automated and technologically advanced manufacturing processes. While we may have less people in the manufacturing workforce, we will have higher paying jobs. Because we are competitive in the global markets, our production in this country will expand and will result in an increase in the higher paying more skill-based jobs. Rick Romell from the Milwaukee Journal Sentinel writes:[27]

ManpowerGroup, after overseeing a huge survey of employers worldwide on the impact of automation, has a calming message for workers worried about the digitized future: Don't fear the robot. Among the employers polled, only 10% told ManpowerGroup that automation would prompt headcount reductions over the next two years, the Milwaukee-based staffing company said Friday. Twice that share — 20% — said they expect to add workers because of increased use of digital technology. Two-thirds of the employers, meanwhile, said automation would have no effect on their staffing levels over the next two years. The findings come from a survey conducted in October of 19,718 employers across forty-two countries and six industry sectors. The results vary by country and occupational group. U.S. employers were among the most optimistic, with 25% saying automation will increase their headcounts in the near term.

TOP FIVE STATES FOR MANUFACTURING

The December 2017 article in Manufacturing Talk Radio highlights the fact that the culture from state to state varies. While some states' cultures embrace manufacturing companies and employment, others do not. Below are the top five states that are making strides in making advanced manufacturing landscape within their own economic footprint.[28]

GEORGIA ranks high on best places in America to work in manufacturing. According to Forbes, the share of manufacturing jobs in the local economy came in at 6.1% and as the U.S. manufacturing industry continues to expand, manufacturing employment is expected to rise along with it. The Bureau of Labor Statistics found that in October in 2017 there were over 387,000 individuals working in the Georgia manufacturing industry. This makes an impressive chunk of the over 12.3 million Americans employed in the industry.

INDIANA. From November of 2013 through November of 2014, manufacturing employment grew at the impressive rate of 4.5%. Employing over 531,000 manufacturers in October of 2017, the employment rate has been on a steady incline since January 2010. As the U.S. becomes a more attractive country to set up a manufacturing business, jobs growth is expected to grow and Indiana will continue to be a manufacturing leader.

MICHIGAN. Known well for their automotive manufacturing industry, that's not the only thing that put Michigan on the map as an industry powerhouse. Fabricated metal products, machinery, and food and beverage manufacturing are all important sectors that help drive the local economy forward. With over 603,000 manufacturers employed in the state, Michigan earned their spot as one of the best manufacturing states. With so much opportunity, Michigan is sure to remain a manufacturing leader for years to come.

TENNESSEE. Ranking in at number four for best states for business was the great state of Tennessee. The state's total output from manufacturing was $48.39 billion in 2014 with an average annual compensation of $66,431. Tennessee employs around 345,000 manufacturers and has been growing since 2010. Across the country, manufacturing is steadily moving forward and Tennessee is poised to embrace this growth and capitalize on the expanding industry. It'll be exciting to see how Tennessee's manufacturing industry performs as time goes on.

SOUTH CAROLINA. The state employs over 247,000 manufacturers and employment has been on a steep incline since January of 2010. Manufacturing employment in the state has almost completely recovered since the dramatic decline experienced in 2008. This incredible recovery earned South Carolina a spot on the list for the manufacturing states and with American manufacturing expected to grow, employment might even surpass pre-2008 levels.

These five states employ over 2,113,000 manufacturers and are on a path of continued growth. It's exciting to see states known for their manufacturing industry once again begin to thrive. The future looks bright for the industry as a whole, but these five states are the ones to keep a close eye on, especially if someone is looking to begin their manufacturing career.

A CONVERSATION WITH TONY SCHUMACHER
-NHRA CHAMPION
WHICH COMES FIRST: MADE IN AMERICA
OR BUY IN AMERICA?

About 2000, I had been to Mother's Day brunch with my extended family when my sister Amy mentioned that a friend from work was moving into our subdivision. She thought the husband did something with racing, but wasn't totally sure. I mentioned that I would stop in and say hello. Not long after the family had moved in and was settled, I knocked on the door and introduced myself. As it turns out, they were a young couple named Schumacher. Tony was a drag racer. Over the years, we became very good friends, and it was awesome to see Tony grow and become so accomplished as now an eight-time NHRA (National Hot Rod Association) champion.[29] (Tony recently set a new speed record of 336 MPH).[30] Besides owning Don Schumacher Racing, (DSR) his family (father) also owns Schumacher Electric. Schumacher Electric is known as a first class battery charger manufacturer. Tony is ideally suited to talk about so many topics in this book. He became the winningest Top Fuel driver in NHRA history.

Now, if Americans, as buyers, would walk into a store and only buy the stuff made in America, we'd be great. We would be wonderful. But they go in and they want the cheapest price too. When I hear people go, "You guys make your product outside the U.S." Really, you buy product outside the U.S.? Stop buying it outside the U.S., and we can start making it in the U.S. It costs more to hire an American than it does to hire someone in China. We still save a ton of money, even with the shipping and all the containers, the taxes, everything coming in,. As buyers, we need to make the conscious choice that it's here that we want to be part of. Stop being so, "We're trying to save money." It comes full circle. If we spend the money in the U.S., we all get paid more, everything goes up. The value of everything we do is good. I'm not making anything up. We can Make America Great Again, we just got to do it right. We got to start being responsible as a buyer, buying a product right here, so people who own companies can start making stuff here, because we want to, and guess what? If I want to go overseas and make stuff, don't buy it. Maybe you'll force me to do it. I think it's important that we take a little responsibility as the buyer.

While speaking to engineers at Boeing, I told this story. If I was going to get in my race car that goes 330 miles per hour, and I'm about to start it, and my crew chief leans over and goes, "I've got great news for you. This car's built in China and Mexico. I just saved your dad a ton of money." My response would be, "I have great news for you. You can drive. I'm not getting in that car." The place went nuts.

On a very bright note, there are some things that are starting to make a difference in the U.S. The STEM and Project Lead The Way initiatives are making their way into the high schools. STEM is an acronym for Science, Technology, Engineering and Math. Project Lead The Way allows for high school youth to learn with their hands and makes the connection from book knowledge to project-based learning. First robotics and other robotic team competitions teach young people decision-making skills as a means of developing the much-needed troubleshooting tasks that will be needed in any manufacturing field or career. Years ago (around 1994 time frame), I followed a lead by a fellow machine tool sales professional Tim Doran to do presentations at local high schools. This would encourage students to attend our trade show named IMTS (International Manufacturing Technology Show) that is held every two years in downtown Chicago. This is a practice that opens an entire new world to the youth of this country. I try to explain that IMTS is like the Disney World of the machine tool and manufacturing industries. For so many years, attendees had to be over eighteen, which of course did very little to attract the next generation of workers.

Having said that - the biggest threat we have is the lack of skilled workers that exists.

As The Reshoring Movement progresses, the new administration removes the regulations, taxes and other hindrances that impede manufacturing companies from competing in a world market. We MUST address the lack of employees to allow the manufacturing renaissance to bring prosperity back to the USA and the middle class of America. This means more training, opportunities and programs.

I hope that this chapter has enlightened you in many ways. My hope is that you have a newfound appreciation for the importance and significance of manufacturing and our country's place in the global perspective. Automation is a means to building our economy and needing even more skilled workers, not less. Chapter three will present those who hope to reinforce the CHAMPION Now!® messaging by changing perceptions through plant tours, podcasts, talk radio, and

TV programs. They are my heroes. Collectively, we will be the key for all of us to be able to build tomorrow's U.S. skilled workforce.

CHANGING THE PERCEPTION OF MANUFACTURING:
America's Greatest Business Opportunity

"They keep going down this road like sheep to a slaughter, man. Everybody told me I couldn't have a shop and that American manufacturing was dead and I was like –
'I'm different.'"

– Titan Gilroy, CEO Titan CNC Academy

Manufacturers have done a poor job of marketing their careers. No longer can we sit back and complain. Rather, we need a coordinated campaign to change the culture of this country by changing the perception of manufacturing and other limiting beliefs.

CONTINUED CONVERSATION WITH DON DUPREE
"A PIECE OF THE GAME"
FOUR TIME EMMY AWARD WINNER

I think you're on the right path. Everybody seems to be looking for a job. I see jobs in Jacksonville, our old home town. I see Amazon opening up a plant there ,and you have thousands of workers lined up out there applying for jobs, even seasonal jobs. It's a pretty powerful incentive. Those aren't well-paying jobs. It's interesting for people. "Hey, I'm Joe. I graduated from high school. I didn't know what I was going to do. I didn't want to go into debt with a couple hundred thousand dollars in college loans. I looked at something you told me about, you know, this training. I went through a training period for six months. Now I've got a job that's high paying. I don't owe any college loans. I didn't spend four years getting it. I'm in the workforce right now with benefits." That's pretty sexy, you know?

You've got a good story to tell. I mean, that's the first thing. How you make that content interesting is the second step. You've got something that people are going to respond to and want to find out more. I think your way of distributing it, like we're talking about on social media, is perfect for what you're doing.

What a great thing for them and for you, to make it "sexy" if you will, the manufacturing thing, but also to get things out there. I think it's great. I think you've got a great story to tell. It's going to be a challenge getting it out there, but I think we're at the perfect time to get the word out. I like the book idea.

MAINSTREAM TV COMMERCIALS

There is a commercial from MB Financial Bank. With the "You know your business" tag line, The lead character is walking through his machine shop stopping to explain various aspects of his company to the banker. While the shop owner is rattling off various aspects of his CNC (computerized numerically controlled) lathe, his coordinate measuring machine, and his press brakes— the MB Financial banker does not say a word in the entire thirty second spot. This type of commercial speaks volumes as to our culture when no one in the general populous understands manufacturing. Case in point is the manufacturing community needs to do just that. SELL the public on what you do. How many times do we meet in social settings and people ask you what do you do? Your response is something that you are never quite sure how to deliver. Will they have some clue as to what I am talking about? Or will they politely seem interested with the deer in the headlights glare, while thinking to themselves the lesser glamourous perception that we have grown all too familiar with. Manufacturing Day® is a perfect opportunity to counteract that. Let's be loud and proud of what we do and how we make the United States an economic powerhouse. Gillette has also started to lead the way about the pride of making quality products in America. Some of their latest commercials talk about the company competing globally with American workers making a quality product.

CNC ROCKS!™

We have already covered the fact of how important manufacturing is to this country. What we haven't covered is how absolutely awesome today's manufacturing technology is! I sincerely believe that given the exposure to CNC—Computer Numerically Controlled machine tools, today's youth will find manufacturing to be exciting and fascinating. CNC is where it's at. The next generation is and has been brought up in a digital environment. They text rather than pick up a phone. They play computer games rather than go outside and play a pickup game of soccer, football, baseball or hockey (unless of course it is organized ball). They start using their iPhone around the age of eight or nine. My point is that with this culture, why aren't we having kids lined up for careers in manufacturing? They have all the innate capabilities from a very young age.

The answer is back to the CHAMPION Now!® message. The perception is that they can't make good money. It is not a safe or clean environment. It is not for smart students or an honorable profession. The facts are that the manufacturing sector averages about $82,023 with benefits a year, while all non manufacturing industry jobs average only $64,609 with benefits per year.[1] We need to change the perceptions not only in the young people, but also the parents, grandparents, the media and the guidance counselors. That is the difficult task. This is something that will take time and not come without a price.

TV Shows, commercials, news reports, an active approach to plant tours, a program to educate guidance counselors of choices available. My hope and vision with CHAMPION Now!® is to have a consolidated marketing campaign across the country. If just one sitcom could portray a character in a positive image, with a popular actor making the manufacturing environment cool or accepted, this would be the start.

The term CNC Rocks™ is a slogan to convey that manufacturing is a great place to be. It is exciting. The CNC portion of the industry makes it pertinent to the youth. I thought of a lot of other slogans, but this one spoke to me (and wasn't already trademarked by someone else) in a way that makes it hip to the youth. What is CNC? – will be an obvious question. That will be where it all starts. Once they ask, then it becomes a cascade of exiting plant visits, or videos, or YouTube hits, or all of the above and more. That is what we need. We need to intrigue the young people, let them find interest, and then not talk them out of it, but instead allow them to find a passion in it. It has worked for millions of people like me and many families like ours.

It is time to make manufacturing and the other "skilled trades" cool! By educating young people what CNC is and stating that it ROCKS, maybe that can get their curiosity to peak and have them start asking questions. I have stated to many young people that if you look into this and you decide this is not for you, that is OK. What is not ok is for us in the manufacturing community to not make certain that young people even know that these careers are there to begin with. Good paying, exciting, challenging positions with computerized and automated equipment most, of which today's youth do not even know about—that is what my quest is all about! CNC ROCKS!™

We don't want young people to lose sight of following their passion and what they love to do is MOST important. If they enjoy making things, designing things, and working with their hands, then they are naturals for the manufacturing world. Many manufacturing companies will pay their employees to go to school to further their degree. They will pay for the employee's classes at night while they work during the day. Manufacturing companies have gotten very good at mentoring young people as they enter into their workforce. Many of them are 50+ years old. They have a genuine concern for who is going to learn what they already know. They want to pass on their skill sets, talents and knowledge to the next generation.

THE DISCONNECT BETWEEN PERCEPTION & FACTS

While Manufacturing is filled with high paying jobs, people aren't joining the field

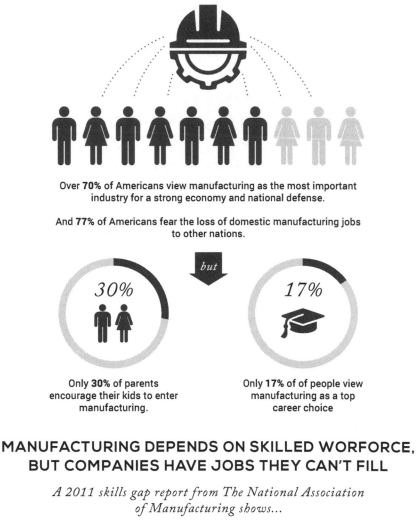

Over **70%** of Americans view manufacturing as the most important industry for a strong economy and national defense.

And **77%** of Americans fear the loss of domestic manufacturing jobs to other nations.

but

30%

Only **30%** of parents encourage their kids to enter manufacturing.

17%

Only **17%** of of people view manufacturing as a top career choice

MANUFACTURING DEPENDS ON SKILLED WORFORCE, BUT COMPANIES HAVE JOBS THEY CAN'T FILL

A 2011 skills gap report from The National Association of Manufacturing shows...

67% of manufacturers reported a moderate to severe shortage of available, qualified workers.

By 2030, **77%** of skilled baby boomers will have left the workforce.

Source: The National Association of Manufacturing

THE MANUFACTURING DAY® MISSION – CHANGING PERCEPTIONS

WHO'S BEHIND MANUFACTURING DAY®?

In May 2012, The following organizations met to form an alliance to change perceptions of manufacturing careers in the United States. Iverson & Company hosted the meeting at our office. Gardner Publications, Sandvik Coromant, Association of Manufacturing Excellence, The Manufacturing Institute, The Reshoring Initiative, Technology & Manufacturing Association, Chicago Manufacturing Renaissance Council, Fabricators & Manufacturers Association (FMA), Foundry Educational Foundation, Cast Metals Institute, Sound Quest Productions, Automation Federation, National Center for Manufacturing Education, Society of Manufacturing Engineering, SME Education Foundation, Edge Factor and CHAMPION Now!® During this meeting, I suggested a day that we could promote manufacturing. Pat Lee from FMA spoke up and said, "We are working on that right now!" Well then we don't need two groups doing the same thing. Let's see where this goes. Manufacturing Day® was born shortly thereafter. Over the years, gradually more and more companies open their doors to SHOW and TELL what they do. Thousands of people each and every year learn that there is something to this manufacturing thing!

Created by Founding Partner Fabricators and Manufacturers Association International in 2012, Manufacturing Day® has enjoyed support from many organizations aligned with its mission of positively changing the public perception of modern manufacturing. Organizations that have played a vital role in working with FMA to successfully grow this national celebration of all things manufacturing include the National Association of Manufacturers (NAM), the Manufacturing Institute (MI), and the National Institute of Standards and Technology's (NIST) Hollings Manufacturing Extension Partnership (MEP).[2]

Manufacturing Day® addresses common misperceptions about manufacturing by giving manufacturers an opportunity to open their doors and show, in a coordinated effort, what manufacturing is — and what it isn't. By working together during and after Manufacturing Day®, manufacturer's will begin to address the skilled labor shortage they face, connect with future generations, take charge of the public image of manufacturing, and ensure the ongoing prosperity of the whole industry.

Manufacturing Day® is designed to amplify the voice of individual manufacturers and coordinate a collective chorus of manufacturers with common concerns and challenges. The rallying point for a growing mass movement, Manufacturing Day® empowers manufacturers to come together to address their collective challenges so they can help their communities and future generations thrive.[3]

Here's the eyebrow-raising section for business, politicians and economic leaders. We will measure the importance of manufacturing. Manufacturing employment has gone down. Why? Outsourcing, increase in productivity. You can't just measure manufacturing effectiveness or performance by employment. Measure it by a function of GDP. Manufacturing in the U.S. is the eighth largest economy in the world.[4] So how can some politicians and our general culture say that manufacturing is dead?

6 MYTHS ABOUT U.S. MANUFACTURING
DEBUNKED!

There are a lot of myths out there about U.S. manufacturing. Maybe they come from misinterpreting the news, politicians making misinformed promises, or just counting the number of "Made in China" stickers we see. No matter how these misconceptions about U.S. manufacturing have come about, the fact is these myths are truly detrimental to its continued growth. Please be sure to read on after our myth-busting list to learn why it's critical for perceptions about U.S. manufacturing to change and what industry leaders are doing to get this message out. The website www.Marketing4Manufacturers.com says it best:[5]

Myth 1: Manufacturing jobs in the U.S. are disappearing, and they don't pay well.

Fact: In 2015, millions of manufacturing employees made on average $81,289/

year, a salary high above the U.S. average of $50,756.[4] In addition, nearly 3.5 million more manufacturing jobs are expected to become available in the next ten years.[6]

Myth 2: U.S. Manufacturing isn't a big part of our total economy any more.

Fact: In 2016, U.S. Manufacturing contributed $2.18 trillion to the U.S. economy and accounted for 9% of the total workforce. Additionally, manufacturing accounted for over 11% of GDP in the economy.[7] For every $1.00 spent in manufacturing, another $1.81 is added to the economy. That is the highest multiplier effect of any economic sector. In addition, for every one worker in manufacturing, there are another four employees hired elsewhere.

Myth 3: Manufacturing jobs are disappearing and the ones that exist don't require much skill.

Fact: There are an estimated 350,000 manufacturing jobs unfilled today. [8] More than 80% of manufacturing companies have reported shortages in skilled production workers from entry level on up. Since, increasingly, machines are computer controlled the jobs that need filled are for programmers, operators, and maintenance workers. Highly skilled maintenance workers can earn more than $100k per year.[9]

Myth 4: Free trade agreements hurt U.S. manufacturing.

Fact: About half of all manufactured goods exported from the U.S. went to countries with which we had free trade agreements. U.S. manufacturers in 2015 had a $412.7 billion surplus with countries with which we had free trade agreements versus the $639.6 billion deficit they had with countries without agreements.[10]

Myth 5: There aren't many 'small' manufacturing firms in the U.S. any more.

Fact: Over 98% of manufacturing firms in the U.S. are small. In other words, there are 248,152 manufacturing firms in the U.S. with fewer than 500 employees. Furthermore, 75% of these small manufacturing firms have less than twenty employees.[11] It's also interesting to note, manufacturers have one of the highest percentages of workers who are eligible for health benefits provided by their employer—92%![12]

Myth 6: U.S. Manufacturers can't compete with cheap, overseas labor.

Fact: Over the past twenty-five years, export of U.S. manufactured products has quadrupled. Also manufacturers fund more than 75% of all private sector research and development in the country, creating more innovation than any other sector. With the very large amount of growth the manufacturing industry has experienced and its focus on innovation, U.S. manufacturers have become more "lean" and automated which is helping them become more competitive in the global market.

BIGGEST THREAT TO THE U.S. MANUFACTURING INDUSTRY

With these myths debunked, it's clear that the U.S. Manufacturing industry is thriving and enjoys the promise of even greater growth and prosperity ahead. In fact, 90% of manufacturing companies recently surveyed feel positive about their own company outlook. The statistics hold the promise of new prosperity, benefiting our global competitiveness, economic growth and communities across the country. But there is one huge threat to the future of U.S. Manufacturing that we MUST ACT NOW to change: there is a large deficit of skilled workers to fill open positions. 350,000 positions unfilled to be exact. And this number is expected to increase in the future!

INTERNSHIP PROGRAMS
THE KEY THAT OPENS UP THE DOOR TO OPPORTUNITIES

I was an early adopter in the internship program. I had students from Austin Polytech, District 214 in Chicago suburbs, some brought to me by NSERVE – an organization that works with nine high schools in Suburban Chicago. Martha Eldredge-Starke is the Executive Director for NSERVE. This organization does a great job, expanding their offerings to include a computer based mentoring program called "Inspire Your Future" through "Career Cruising". Julianne Arvizu is their Coordinator who I met when I was working with Oakton Community College on planning their Manufacturing Expo for many years. This Expo was hosted at Oakton, and brought local high school students in to meet with local manufacturers to learn what they do, and careers they offer. Both Martha and Julianne are inspiring in how they help young people everyday in manufacturing career paths.

A CONVERSATION WITH RAND HAAS AND MCIP INTERNSHIP PROGRAM

I got involved with Rand early on, probably four or five years ago, and have been involved with him and his program for about that time. In Rand's first internship group of ten, I took two of the ten and hired both. One still remains with Iverson & Company. The internship program is in general, I think, vital in trying to bring along young people towards manufacturing careers, and in some ways, can be combined with an apprenticeship program – to which I call an ApprInternship (a new term that I coined). I just recently read an article about how an alternative to the conventional apprenticeship model is necessary. I'll go on to say that Rand's program is by far one of the best that I've seen, and I've been exposed to maybe five or six different internship programs. I think there are valid reasons why he was awarded recognition.

The MCIP (Manufacturing Careers Internship Program) program has been recognized as one of the most effective and progressive programs in the United States. It received the innovation award from the State of Illinois for the most innovative workforce program in the state. It also has been followed up with by the Aspen Institute out of Washington DC as an example of one of the top three workforce development programs in the nation.

The boot camp is the reason for the success of this program. The program has over a 70% success rate. These young adults previously had no jobs, no experience, and no skills, and frankly no future. 70% of them now are either employed or have gone back to school. It's a very robust four week boot camp where we combine classroom with practical experience. We found these young adults come out of school with no idea of what's expected of them in their workplace. We call this work readiness training.

There was a group out of Denver, Colorado that did a study of over 1500 companies. They asked the HR departments, "What are the soft skills you look for when you hire and promote?" They took all the answers and came up with seven attributes that every employer was looking for. That is whether it's manufacturing, retail, sales, or healthcare. The seven attributes are: attitude, attendance, ambition, accountability, acceptance, appreciation, and appearance. We also added an eighth one. The eighth A is "ask." Those eight A's have become the basis of our boot camp. We cover these in the first three days of boot camp. Then we go out and we visit anywhere from eight to twelve different companies. Each will ask the owner or the general manager, "What do you look for when you hire or when you promote?" We ask them, "Is attendance important? Is attitude important?" Then we dive into why it is important? We then we go one more step farther. We ask again, "Why is that important for when you get promoted?" These young adults hear the same message repeatedly from different sources for four weeks. At the end they realize, maybe I need to change my attitude or improve my attendance. The ones that do are highly successful. We also teach them other skills that are certifications such as forklift driving. All the interns have a forklift driving certification. They also have a ten hour OSHA (safety) certification.

The importance of these certifications is not only to make them safer as well as give them a nominal skill in forklift driving, but it also helps your self-esteem. Most people don't have these certifications when they're nineteen years old and looking for a job. That gives them an edge. And these young adults realize that they have nothing else to offer right now except work ethic and certifications. We find they're very meaningful, not only to the employer but also to the intern. It helps build their self-confidence that they can achieve something. The other part of the program also includes basic math. We call it shock math. Conflict resolution, when something goes wrong at work, don't just walk out. Talk it over with your management. We talk about financial literacy. You're making money what do you do with it? How to save for the future. How to build credit.

At the end of those four weeks, we do an interview with all the employers. We call it Internship Selection Day, where the employers come in and all the employers interview all the interns. The purpose of this is to help the interns learn how to do a job interview. Prior to the internship selection day, we do mock interviews. We videotape them. We help write their resume and prepare them for the all-important question, "Tell me why I should hire you?" We help them answer that question so they don't draw a blank stare when they are asked that question. These are the reasons that it's successful. We also have a dedicated staff that follows these interns once they get the internship. The staff calls these interns at least once if not two or three times a week. (We) make sure that they're showing up with no issues at home. These young adults have issues that they're dealing with that transcend the workplace: could be poverty, could be they have a child on the way or have children already. Or issues that we're not aware of in the workplace. There's more of a holistic approach to this program, which helps ensure their success in the workplace.

The selection process is similar to a job fair. It's a private job fair for just the employers and the interns in the program. Every intern has about fifteen minutes with each employer. At the end of the selection day, each employer writes down their top three or four interns that they would consider hiring. The interns write down their top three or four employers. This creates buy-in. Afterwards the staff sits down and looks at which interns want to work at which location.

During the four week boot camp, we were also analyzing these young adults—whether or not they were serious or just killing time. We've realized that this intern is not ready for the responsibilities of a full-time job. One thing I really like about the program—we don't just throw them off the bus. We find another bus for them to ride and to help them get on. Whether that's going into culinary skills or going into healthcare. When I developed the program back in 2011, I realized that not everybody has the inclination or aptitude for manufacturing. The problem is until you try it, you don't know if you do or don't. When we first started as a pilot program, we were thinking 25% of young adults would go into manufacturing. We were wrong. Almost 75% of that first class went on to a career in manufacturing. Well we tried again with a different group. Same results. 30% aren't ready. We work with career advisors to help them with other issues that may be preventing them from work: could be transportation, could be don't have a GED or high school degree. It could be issues at home. We work with these young adults on an ongoing basis for a year. They get in the program and are successful. If not, we work with them until they are.

Back to the perception of manufacturing in America. I have some statistics that even though 80% of people say manufacturing is important to the economy, less than a third of Americans in the survey would want to have their child working in manufacturing. That's part of the perception issue. Most people grow up thinking that manufacturing is for dummies. That it's dark and dirty and dead end. No single company or organization can change that perception. The MCIP program is a turnkey program for manufacturers to jump on a successful program. Collectively through these tours, internships, and jobs, they are changing the image of manufacturing.

Employers play a huge role. I have been doing this now since 2011. I am constantly surprised at the commitment manufacturers make to the success of the MCIP. It's amazing. Their goal is to make money for their owners and shareholders. And yet, they're the corporate citizen. They need to help the less fortunate. I have been surprised by how many times that even though the internship may not work, the manufacturers are willing to try it again. I cannot thank the employers enough for their interest in this program. We've had over 450 young adults go through this program. We're working with over 135 companies (across several counties). These young adults are talking to their friends and are getting into the program now. The change is granular. It's slow. One thing we found with dealing with millennial is that you have to become a trusted resource. And the only people millennials trust are others of their own age. By promoting this program and working side by side with CHAMPION Now!®, we can reach more and more young adults. I see the tide turning. I see it in high schools now. The perception of manufacturing is changing. It just takes time.

A CONVERSATION WITH JIM CARR
"MAKING CHIPS" IS WHERE IT IS AT – MARKET IT!

Speaking to the perception of manufacturing, few machine shop owners are marketing savvy enough to tackle such a huge undertaking. During my time at the TMA (Technology Manufacturing Association), both my daughter Lindsay and I crossed paths with a brilliant marketeer who just so happened to own a machine shop – Jim Carr. Jim has moved into the world of podcasting and has moved mountains to get the get the word out about manufacturing opportunities one podcast at a time.

I'm the cohost of a manufacturing podcast called MakingChips. My dad founded a machine shop, CARR Machine & Tool, Inc. in the garage of our Mount Prospect home (in suburban Chicago), in December 1972. I worked part-time for the company through high school, and in my senior year, my parents asked, "Listen, you have an opportunity to work full-time in the family business. You can do an apprenticeship program through the Tool and Die Institute in Park Ridge (now TMA in Schaumburg). If you're not interested in the family business, you can go to a four-year university, and get an education in that capacity." I thought long and hard, and at the ripe old age of eighteen, made the decision that the family business could probably be more lucrative in the long haul. I made my choice.

In November of 2013, I was asked to be interviewed on a Chicago A.M. Radio station because I was doing something uniquely different in the manufacturing space, and their editors thought it was interesting. I was probably the only machine shop in the Chicagoland area using social media to promote my company. It was kind of an out-of-the-box thing. At that same interview, there was another guy by the name of Jason Zenger, (who was the president of ZENGER's Industrial Supply). I had never met him, but CARR was buying industrial tools from them. After the show, about two weeks later, Jason called me, and said, "Hey, Jim, I just got an idea, one of many thousands that come in my head." He said, "Do you listen to podcasts?" I said, "No." He said, "Well, do you know what a podcast is?" I said, "Yes hasn't that media been around for about ten years?" and he said, "Yeah, but, with the advent of 4G technology, and Bluetooth capability in your car, they're really gaining in momentum." He said, "There's really a void. There's nobody relevant doing it in the manufacturing space." He says, "I think you're pretty well-spoken. You've certainly got a lot of years' experience behind you. You have a great network of people. I think that if we get together, it can be successful."

I thought about it for a couple days. The light bulb went off in my head. I had that a-ha moment, much like I did when I decided to delve into social media. I called him back, "Yeah, I'll do it. I'll take the chance," I said, "but it's gotta be really professional," He said, "okay." For one solid year, we planned all the idiosyncrasies, and logistics of putting out a professional podcast.

We're into it over three years now, 151 episodes. We try to get one out every week. Our objective for doing the show was to equip and inspire a manufacturing leader, like a peer-to-peer sharing opportunity. We just wanna be a content-hub resource for industry professionals. We've got a lot of ideas on how to monetize. It's just a lot of work. We're still leading our respective companies.

I really think people should start thinking differently about manufacturing. We all can't be baristas. We cannot be a service-oriented country. We have to make stuff and be that kind of a country to be successful, first and foremost. Manufacturing has gotten a bad rap the last few decades. People's ideas about the industry is that it's dirty, for the undereducated and misfortunate. It couldn't be further from the truth. Our shops are clean, well-organized and air conditioned. The work we do takes craftsmanship, skill and requires a highly technological thought process. And did I mention these are high-paying careers that can support a family for a lifetime?

The biggest thing that I've learned is for a machine-shop owner that's stuck in his business: Get out there, and network. Find a manufacturing association; find a few good people; get together and talk. Open yourself up a little bit. Be a little bit vulnerable, and keep at it. I genuinely believe that you will find success in yourself and in your own business.

A CONVERSATION WITH LEW WEISS OF MANUFACTURING TALK RADIO

Lew Weiss and I are a similar age, but with a full grey beard, round spectacle glasses and long curly gray hair, he and his co-host, Tim Grady, stand out in a crowd in their bright yellow sports coats, black shirts and paisley ties. From podcasts to internet Manufacturing Talk Radio, Lew and Tim are making their mark to change minds about the importance of and the bountiful opportunities in manufacturing. What I find fascinating is that Jim Carr (MakingChips) and Lew Weiss started their broadcasts essentially at the same time in the fall of 2013! Here Lew talks about his ideas on manufacturing.

I've been in manufacturing fifty-something years. I've always done the marketing for my company. I always found that we were over-successful with marketing, to the point that our competitors would always copy us within a two, three, or a four-year period. They are trying to do what we were doing. So we would always change.

I came up with the idea of talk radio. It hit me in the middle of the night in November of 2013. My attitude about that was, well if my competitors want to copy me, good luck to them. This is not easy to do. I contacted my associate, Tim Grady, who's also a cohost on the show. I said, "This hit me in the middle of the night, we gotta do this." Two weeks later, we did our first show. We've been doing the show every week, which is almost five years now.

Nobody is doing what we're doing. The mainstream media, unless it bleeds – it doesn't lead. We just keep presenting stories that you won't hear outside of

Manufacturing Talk Radio.

We have a new show called Women And Manufacturing (www.womenandmfg.com). It's all about what's happening in the manufacturing world in regards to women. It's really an amazing story. Our country is in a serious issue with regards to people retiring—the skills gap. The news isn't getting out. The information isn't getting out to those who need to know it.

We wound up doing, twenty to thirty shows on Manufacturing Talk Radio when we realized that there was a benefit to talking to people of known stature who are working hard at getting the information out. People like the National Institute of Science and Technology, New Jersey Manufacturing Extension Program, The National Association of Manufacturers, Institute of Science, Institute of Supplier Management; All these organizations are working their butts off to get information out. They're getting out as clumps of information. There's no major funnel to get the information out.

We are actually presenting information that they had no knowledge of. You're never going to know about it because no one's going to put it out there. CNN and MSNBC and Fox and Friends, and all the rest of them, that's not what they're into. All of the talking heads, they talk about manufacturing as representing 12% of the economy. That's not true! The real number is 32%. It's the upstream and downstream of organizations that work with and support manufacturing. For example, logistics, air freight, the chemical industry. All of these companies that produce parts and services that support manufacturing are part of the manufacturing economy. It really isn't 12%, it's really 30%, 33%, somewhere around there. That's a huge number. For us to allow organizations and government to ignore it and not treat these problems –our country's in serious trouble.

Tim and I have researched and talked about women and manufacturing as they did in the Second World War. Four million women went to work. A couple months ago, we had a woman by the name of Anna Hess who was an original Rosie the Riveter. She was fifteen years old in 1942, forty-three. She worked in a tire manufacturing plant in the Midwest. These are the people that helped support manufacturing then. Women only represent 27% of the workforce in manufacturing today. Meanwhile, they represent 51% of the population. We have a real serious disconnect.

We're doing our thing to try and make the connection. That's why we started last year with Women And Manufacturing. We've got five hosts. Women who have reached a certain level of competence, education and knowledge about manufacturing. They all have guests on the show. We're listening to all these stories about how these women have gotten to reach this level of management. We are planning a couple of show series talking about women who may not have reached the C suite level but may have reached managerial level on the floor, managerial level in the office by understanding the things they need to do to help manufacturing.

A CONVERSATION WITH TITAN GILROY
TITAN TV AND TITAN MANUFACTURING

I have been fascinated with a new force who has recently come on the scene of manufacturing. His story is quite special. He took the road less traveled and came to be a significant force in manufacturing education. Titan Gilroy is his name, and he has started bridging the gap to television and technical education. Like Jim Carr, he also owns his own CNC machine shop.

My name's Titan Gilroy. I'm an American manufacturer. I started my company in 2005, after ten years of working for different shops. I entered the industry without even knowing what a machine shop was. I started working on parts for Siemens and NASA in a prototype type atmosphere in the Bay Area. I've always loved making things. Ever since I was a kid, I was very creative. I didn't have money. I built my own toys. I built my own surfboards in Hawaii, and I painted.

I took to manufacturing naturally. It's a cool story. I owned my own machine shop at thirty five. I met with my grandmother for the first time. She showed me a picture of my grandfather who was a machine shop foreman at Boeing. Later, I met my other grandfather for the first time—he was a machinist for Chrysler. I walked into the trade for nine bucks an hour and never knew the heritage of great machinists before me.

I grew up in Hawaii fighting and being the only white kid in my neighborhood. I went from scrapping on the streets and getting in hundreds of fights because I didn't back down when people picked on me. I ended up being one of the top fighters in the United States and fighting for Top Rank boxing. I ended up taking all of my aggression and my competitive nature and putting that into the CNC machines that I was working on. Other people would sit there and push a button. I was more curious 'How fast can this thing go? How can we make the finishes better? How can we run more parts and compete?'

If you solve problems for the right people and make them money, you can be incredibly successful. It's about striving for perfection. You have to be proud of the quality that you are producing and blow people's minds. I want my customers to be blown away by my products. I get comments all the time, that the parts look like jewelry! That's what it's all about, having pride in your workmanship.

After solving huge problems in the industry and making a name for myself and my company, the recession in 2008 took it all away. I almost lost my shop. Every day for a year, I sat on a bridge praying for it to be saved. Ultimately, I saved my shop but something was different. My perspective changed and had a heart for the industry after seeing 50,000 shops close. That's when I knew I had to do something and developed the idea for my own TV Show (TITANS of CNC), which is the only national television show featuring CNC Machining. I wanted to inspire people with the possibilities that manufacturing had to offer. I wanted to tell the stories of great companies and open the doors to an industry that no one knows anything about. It still plays every Monday night on MAVTV, and it's continually growing.

Even with the TV Show, I knew that more had to be done, and I dedicated myself to CNC education. In my third season, the state of California asked me to develop a CNC program inside of San Quentin Prison, and we documented the entire process on the show. This led me to create my own curriculum for CNC machining and create true change in the industry that is desperately needed. It's been education ever since. I haven't lost faith in the industry, but I've learned to harness my attention and focus it on creating things that matter.

After the third season of the TV show ended, I fully dedicated myself to creating a CNC curriculum and created a revolutionary platform called TITANS of CNC: Academy. This is the most advanced CNC curriculum ever developed. The bottom line is, there is a real problem in this trade right now. We are dealing with one of the greatest skills gaps ever seen, and there is no curriculum to support the training that students need. Currently, students aren't learning enough. They aren't learning through repetition. They only make a couple of parts as individuals and spend most of their time working in teams I've looked at the problem, and they are simply not making enough parts to be successful. Not only that, they're not running the types of parts that matter to the industry that will hire them.

And the advanced classes are simply not advanced. They're still pushing the basic philosophies that were taught twenty to thirty years ago. That's why I put out so many videos and so much content because instead of talking about it, I'm doing it.

Why is our curriculum different? We make what was once complicated, easy. We embrace technology and teach how to use it efficiently. We have tutorials that take you from the first step of designing a part to holding the same finished part in your hand. We've standardized materials and standardized tooling. We've done things in a way that progresses students rapidly. They learn through repetition. Everyone is picking it up. We've got six-year-old kids, thirteen-year-old kids. We've got students all over the world, in 170 countries—all in just one year of me launching the academy. Now shops are picking it up and advancing their companies, making payroll, and buying machines. It's all changing, but one of the reasons it's all changing is because I buckled down and started working and doing what I knew I needed to do.

One thing that I realized was, 'Wait, there's no national curriculum.' There are people trying to make money off of this type of training. This is actually a big problem. Everybody talks the talk, but who sees through all the bull? If you're a principal or a dean of a school and you want to have an education program that you want to make super elite, what do you do? The truth is you have to go find somebody who's great at manufacturing and is willing to teach for the money that you're willing to pay. Nobody is teaching the teachers. Nobody is saying, hey you're doing it wrong. Everybody's too scared to say what needs to be said.

I did not build this platform out of necessity. Necessity, meaning I need money. I'm doing what I'm doing because I'm sick of seeing what has happened to our country and to our kids. We're not competing. We're not teaching our kids to compete. I've been to all these schools. I haven't seen a single one where I thought, 'Man, this is awesome!' These are beautiful machines, but the programs are horrible.

I think that going to college is amazing. What do you want to do? What are your specific talents? If you have the resources and the time and you can do it and you can pull it off and it's for a worthy cause and there's a need, then by all means: Go For It! Going to college is awesome, but not everybody has the same opportunity. There are people like myself who didn't have the resources, the time, or the opportunity. There are so many people like me who are creative and love to work with their hands. Manufacturing opens the doors. If I can figure out and solve some of these problems, I can actually make good money.

My curriculum is free, and it's the most advanced. We now have over 40,000 users signed up for the Academy. 2500 teachers, 3500 businesses, and almost 20,000 machinists have signed up for it. It has become the glue for in-house training. Machine shops have created schools within their facilities using my free education system. They meet in their spare time: on a Monday night, on a Friday evening, or on a Saturday morning and teach the trade to the people in their company. 90% of them open their doors to the public, too. Learn the trade using my curriculum and it's a free education man! These shops then turn around and hire the very people they've trained when they need to. They're learning in weeks what in schools would've taken years.

The next step for TITANS of CNC: Academy is to create live workshops and teach our teachers and our industry leaders how to use our curriculum. Within three days, you're machining your own part. I know this stuff like the back of my hand. I've really looked at it and thrown out all the garbage. Treat me like I'm your wise uncle. Follow me.

I teach common sense principles from a place of understanding. I know what I know after over twenty years of working for the biggest and largest companies. I've looked at teaching as if I was going to teach my own children, I and teach them the quickest and most effective way to get go from A to Z.

Everybody else out there is stuck preparing students for months and months and even years. I'm saying this is actually pretty damn easy. I understand that it may be complicated to some people, but I'm not going to say that to your face. I've taken the difficult and simplified it. I leave you a little trail of crumbs. Follow the crumbs. Don't think about it and you're going to have success.

Everybody told me I couldn't have a shop, that American manufacturing was dead. But, I'm different. I didn't learn from textbooks; I learned by doing and through putting in the work. Sure, I have a different way of doing things. I have a different way of selling my company. Was it hard? Yeah, but look. Now I'm talking to you. We have a TV show. We make crazy aerospace parts. I see the current state of education and, as a leader in manufacturing, I'm bewildered at the fact that nobody's ever fixed it or solved it. They keep going down this road like sheep to a slaughter, Man! Nobody's changing. I'm, like, 'What the heck is going on?' I dedicated myself to education to make a difference, Man! If you want to step into this trade and do something big, then surround yourself with talented people and put in the work. Make it happen, Man!

Titan Gilroy is the CEO of TITANS of CNC and an elite Aerospace manufacturing facility and the creator of the television series 'TITANS of CNC' and TITANS of CNC Academy. Learn more at www.academy.titansofcnc.com

You have read in this chapter about Champions trying to change perceptions about manufacturing in many ways. Marketing careers and opportunities in the manufacturing sector is what each is doing. But this will also require a change in culture. There is an alarming shortage in manufacturing. and we need to act now to address it! Companies are not being able to grow because they do not have the skilled workers needed. Students have amassed debt that will not allow them to have a family, buy a home and live a comfortable life. Adults struggle to find a career that allows them to be productive and happy members of society. Believe it or not, though, there is an even bigger threat to finding America's Champion—bigger than the economy, bigger than any of us. In fact, the future of thousands of young people, as we will soon see in the next chapter, depends on solving this problem.

TALENT VOID: IN THE SHOP AND THE CLASSROOM

"You need to figure out what they're doing (local manufacturers), what they need, and understand and speak their language so that you can have a good relationship with the manufacturers and you can be the liaison to future opportunities for your students."

– Dr. Laz Lopez, Associate Superintendent for Teaching and Learning at HS Dist 214

We have an epidemic in this country and void/disconnect between what is out there and what we need. If this continues, there will be millions of open jobs and millions unemployed. One of my recent interviews was Lew Weiss the host of manufacturing talk radio. He spoke to the numbers of how the time is now to make an impact, and the numbers don't lie—do the math!

A CONVERSATION WITH LEW WEISS
THE NUMBERS DON'T LIE

I guess I just would like to reiterate that if we don't fix this now, we're in a deep, deep issue. Problems and issues: because right now we have 700,000 vacant manufacturing jobs in this country. Ten thousand people a day are retiring, 4,000 people coming into the new workforce. What're we going to do? Because in ten years from now, if we don't fix this, that 700,000 is going to be 2.5 million. Here, we talk about the lack of a skilled workforce and the data behind it. We see the talent in the industry retiring with the baby boomers and the void created—the seat at the table which is now empty. The drop-off is more dramatic than created by overseas outsourcing (which are jobs that are leaving).

TIM COOK, APPLE CEO
CAN WE MOVE MANUFACTURING TO THE U.S.?

On December 20, 2015 my wife Kathy and I sat watching the 60 Minutes TV show in our living room, with an interview with Tim Cook CEO of Apple. As we watched the show, it became obvious to both of us that Tim Cook was conveying the problem that I have been trying to write about and preach to the general public and the educators.

"Most Americans would be surprised to know that many Apple products are manufactured by over 1 million Chinese workers at factories of Apple subcontractors

including its largest, FoxConn. Yet Tim Cook states that China's vast cheap labor force is not the primary reason for manufacturing there!" We will talk more about FoxConn in Chapter Eight. Here is an excerpt from his interview:

Interviewer Charlie Rose asks: "So if it is not wages − what is it?"

Tim Cook responds: "Its skill!"

Rose: "Skills?"

Tim Cook: "It's skill."

Rose: "They have more skill than American workers? They have more skill than German workers?"

Tim Cook: "Now hold on . . . let me be clear . . . China put an enormous focus on manufacturing, in what we would call—you and I would call—vocational skills. The U.S., over time, began to stop having their vocational skills. You can take every tool and die maker in the U.S. and put them in the room we are sitting in. In China, you would have to have multiple football fields . . . because they taught those skills . . . It was a focus of theirs, it is a focus of their educational system, and that is reality." [1]

As these comments came out of Tim Cook's mouth, I looked over to Kathy and said, "Yes! That is exactly what I have been saying!" Kathy nodded and said, "Yeah this is exactly your message." I continued by saying, "I was just telling Greg Wasson this during our breakfast meeting the other day!"

Greg and I talked shortly thereafter about the program. "Greg did you see the 60 Minutes episode with Tim Cook from Apple?" Greg immediately responded, "I did, and when he started talking about the tool and die makers, I looked at Kim and told her that this was exactly what you were telling me at breakfast!"

How fascinating that here our two children married that previous June, and both of us were watching the same TV program and came to the same conclusion about what I think is America's Greatest Champion!

A CONVERSATION WITH BRYAN ALBRECHT
PRESIDENT OF GATEWAY TECHNICAL COLLEGE
"INCLUDING ADVANCED MATH AND SCIENCE"

With the need for more skilled workers, technical colleges are tooling up for the need. Bryan Albrecht is President of Gateway Technical College in SE Wisconsin. Over the years of calling on customers, one customer in Racine was having lunch with me and said, "Terry, you have to meet a friend of mine! He is very dynamic, and I know the two of you have a great deal in common." The person he referred to that day was Bryan. We eventually met and Bryan ended up nominating me for a board position with the CTE Education Foundation that I served on for a year. Bryan speaks to their focus to not only train skilled technicians, but also technicians who have the problem solving elements as well.

In today's job market, technicians must have both technical and knowledge skills. At Gateway, we are ramping up our requirements for technical training to include advanced math and science to provide students with the analytical skills to solve complex problems and troubleshoot technical challenges. Industry 4.0 is driving a curriculum change that has an expanded platform built on the integration of information technology and mechatronics. Advanced manufacturing must be adept in both the physical and cyber worlds to be successful.

HARPER COLLEGE – SPEAKERS UNITE
A CONVERSATION WITH NICOLE MARTIN
GENERATIONAL AGILITY IS THE OPPORTUNE WORD

In 2014, I was asked to be a keynote speaker at Harper College, which had just undergone a rebirth of its manufacturing program through the leadership of President Ken Ender, Vice President Maria Coons and Assistant Professor and Program Coordinator Kurt Billsten. Another speaker there who I was very impressed with was Nicole Martin. I found Nicole charismatic and passionate about the lack of skilled talent in the workforce. She founded her own consulting company. Nicole interviewed me for her book, The Talent Emergency® and her internet program "HR in The Fast Lane."

My book was something that I never imagined would happen, but after meeting with hundreds of CEOs one-on-one over a period of years, it became evident that I was having the same conversation again and again. Often, we take our own knowledge for granted and assume, "Well, everyone in my industry must know this." After a few hundred conversations I realized, "Gosh, maybe they don't know this."

I find that businesses need to change their lens and look at the untapped talent pools and not necessarily look for "qualified" talent. They need to redefine what talent is for them. Women remain a huge opportunity. The reality is that businesses are facing a multicultural demographic in terms of talent pool. Some of the minority pools or different demographic frameworks have not been afforded opportunities or been brought into different industry. There's tremendous opportunity for industry to go specifically after those untapped talent pools when facing the talent gap in the domestic United States.

Every business manager of the future has to have generational agility. Generational agility is the competency which allows the manager the ability to work with an understanding of people and in the advocacy for people from different generations (work ethic, desire, values, framework for rewards and motivations). All of these factors of performance achievement have distinct expectations with respect to each generation. This requires businesses to prepare for what this is and to be agile and comfortable in it. To also know how to treat people uniquely as an individual without comparatively creating disparate aspects between groups, or people.

With the current demographic talent pool, the younger workforce is being put in positions of greater responsibility sooner than they've had opportunity to adapt and be trained for it. That's a result, again, of the Talent Emergency®. The problem isn't just the lack of soft skills. Rather, it is that we're not training people or setting them up for success properly.

Right now, big businesses are competing by giving full benefits to part-timers, or paying for college tuition while they work part-time. Benefits that used to be more mainstream for full-timers are now available to part-timers. That's the new norm. We're living through something right now that is history in the making. A paradigm shift. We're all experiencing something that is playing out before our eyes, and wherever that falls is yet to be revealed.

REQUESTS FROM THE FIELD
"FIND ME AN OPERATOR, AND I WILL BUY A MACHINE!"

Just like Nicole telling owners the same thing over and over for decades, I called on shops that complained about not being able to find good workers or enough of them with the skill set needed to run their companies and manufacturing effectively. They would tell me the same thing over and over about the lack of talented workers. Over the years, it has become increasingly more difficult to field enough talent to run production in most of my customers' companies. The more progressive companies would automate as much as possible in order to meet the needs of their production with as little manpower as possible.

As years progressed, it was pointed out to me that our builder believed in forming relationships with schools. I sat with another salesman at the IMTS listening to how to embark on such a task. As someone who has mentored and coached hundreds of young people, I was encouraged and thought that this was something that going forward, I had to pursue.

My first effort was with Carl Sandburg College in Galesburg, Illinois in 2001. We ended up selling a CNC Lathe and Vertical Machining Center. There was a second project this time in Wisconsin that involved Moraine Park Technical

College and their Advanced Manufacturing Technology Center. We traveled out to visit our manufacturer we represented at the time and even some of the technical colleges that they had formed relationships in Lancaster County, PA. The third and most notable project was Milwaukee Area Technical College. I had called on them for many years. I was always impressed with MATC in the early 1980s when they started a full CNC Program. It was then that I learned what a special place MATC was when it came to providing the learning tools for advanced manufacturing. They were years—possibly even decades—ahead of their time. Around the year 2002, they approached me for a redesign of their entire machine tool program.

During that time, I was really digging into the educational section of our industry. My relationship with MATC was good, and I was finding myself getting more and more enthused and involved. I soon also formed a relationship with Bryan Albrecht, who invited me to be a guest on his radio show which is run out of Gateway Technical College. We talked about manufacturing and what the IMTS Show was and had to offer to young people. It was a great opportunity but just wet my appetite on what could be possible. I thought there had to be more to convey a message to the general public.

WHEN WILL THE MEDIA CATCH UP WITH THE NEED?

At one point, I decided that a national TV Show was needed. Something more than what "HOW IT'S MADE" can offer. This is a great show that focuses on the making of parts, but doesn't show or tell much about the people making the parts, or keeping the machines running that make the parts. I thought then and still think now that this show is great, but what indication would any young person have that there is a career choice attached to the products shown being manufactured? We will speak to this concept more in detail in the previous chapter with Titan Gilroy.

Whether it is a machinist, programmer, lead man, supervisor, set up man, CAD designer, maintenance mechanic, automation engineer, or many other positions—none were woven into the show in such a way that a young person would say, "Hey, I want to do that!" On the bright side though, many adults and young people find it fascinating how a part goes from raw material to finished assembled working product. I approached the producer (Ellis Brombeck) of the MATC public TV station about a show on manufacturing. The producer was interested and said not to give up on the fact that this was a viable dream. I even commented that it could be a regional show, if not national. The producer said, "No national is what we need, keep driving towards that."

In 2009, I attended an industry open house in Schaumburg, Illinois. At that event, I met Pete Zelinski at dinner. He is the Editor in Chief at Modern Machine Shop. I mentioned to him that I had a vision for a national TV show about manufacturing promotion. He sounded very interested and asked that I keep in touch. Although it didn't come up in our discussion, my guess is that there was a conversation either soon before or soon after ours with a talented producer out of Canada by the name of Jeremy Bout.

He ended up starting exactly that very notion, but with films. Modern Machine Shop (Kline Publications) got behind his cause and provided the initial platform for him to show his wares, so to speak. Jeremy had the talent, skill set, creativity, and backing to make this happen. As it turns out Jeremy and I share the same passion and vision.

INSPIRE THEM AND THEY WILL COME – BUT WHO WILL TEACH THEM? A CONVERSATION WITH LAZ LOPEZ
TECHNICAL TEACHERS IS THE VOID WE NEED TO FILL!

With all of this being said, there lies another crisis looming in education. We can inspire all the young people to consider STEM related fields, but if we do not have the technical programs—or even worse not have the instructors to educate them—what then? The only thing worse than not having the young people interested, is to have them on the edge of their seats wanting to know more and then have all that passion die on the vine, with no instructors.

Enter Laz Lopez, Associate Superintendent for Teaching and Learning at High School District 214 in north suburban Illinois. He speaks to our country's inherent problem: a lack of technical teachers. From the first time I met Lopez, I realized he "got" what most do not. I met Laz when I was working with Wheeling High School and their advanced manufacturing program. At the time, he was their principal, and he recognized that with the high concentration of manufacturing companies surrounding his school in a higher than average impoverished economic footprint, that a manufacturing pathway right out of high school was a perfect match.

There is a crisis in talent for all teachers and particularly in technical areas. There are few programs nationally preparing students to teach in technical content areas. There is a need for incentives to encourage technically skilled individuals to serve their communities as teachers. Developing strong technical teachers like those at Wheeling High School, who come out of a traditional public university, requires real world experience. One teacher we hired did not have experience in manufacturing or engineering when he started. You were one of the individuals I reached out to and several others, and said, "I need you to take this new guy, who's young and motivated, and I want him to spend a week with you." He spent four weeks that summer before he started with different partners and had a crash course in manufacturing. He worked the machines for an entire month. He did that to become an expert. I said, "You need to figure out what they're doing, what they need, and understand and speak their language. You'll have a good relationship with the manufacturers and be the liaison to future opportunities for your students."

Schools need employers to collaborate to create externships for our technical teachers so they have the experiences that give them the credentials and credibility to engage in that space.

Laz also comments on the expanding role of guidance counselors.

I would say the role of the counselors has expanded, and it is widely acknowledged that it must now include career counseling. The crisis we have is the lack of professional development needed to provide our counselors the context for them to knowledgeably represent and talk to students about potential career opportunities.

Students in school today will have opportunities that we can't imagine. How do we support counselors to have the tools to help students discover new and developing career.

A CONVERSATION WITH MARK HIBNER
TECHNOLOGY EDUCATION TEACHER AT PALATINE HIGH SCHOOL

Palatine High School was starting a new machining program. They had adopted Project Lead the Way, and there was an instructor there who I got the chance to meet. His name is Mark Hibner, a young man who is the "Pied Piper" at Palatine for the Advanced Manufacturing Program. As time went on, Mark and I found that we both were being invited to the same advisory meetings to make an impact on the educational manufacturing landscape in the Northwest Illinois suburbs.

It's a belief that it's our responsibility as educators to create opportunity for our students. Our students turn out to be like our children in a sense because you develop a great sense of caring for these individuals. You want them to do well and be successful. I was very fortunate to have a teacher that influenced me. Some wrestling coaches became my father figures. I felt like it was out of respect to them. I've had such a great life by becoming an educator. I owe it all to them. I have to pay that back. Also, I work to help others, to help young adults, achieve their goals.

It's just that gratitude that has to be inherent in someone. It has to be that intrinsic motivation to give back like that. A person has to have the sense that they are a servant—a servant to helping others reach their goals and meet those opportunities. I think a big part of it is someone who is enthusiastic for what they do and believes in what they do and what they're teaching. Because I've worked in the manufacturing industry prior to going into education, I just really feel passionate about this discipline. There's so much students can learn from manufacturing. When you look and break down the math and the science and also throw the technology in there, it's pretty amazing. This discipline of manufacturing, it's pretty exciting.

Being really passionate about the content of manufacturing, I feel like it's a patriotic movement. This manufacturing community that's been developed in our area—we get together and we talk and brainstorm ways to create opportunities for kids. It's inspiring. I feel like it's almost like a call of duty. It's a form of patriotism to make the United States a stronger economic force in the global spectrum.

There is an enormous skills gap we have in our educational communities for manufacturing. This is why I feel a sense of patriotism. If we get kids interested in manufacturing, we're going to build a stronger workforce. They're always going to have the opportunity to learn and grow and be challenged. It's about making a stronger community and, as a result, making a stronger country on the basis of working in the world of manufacturing. I was also in military, in the Army National Guard. I feel a great sense of duty to our country.

Mark's comments lead me to think that maybe we should be looking at our ex-Military personnel to become our next advanced manufacturing instructors. Who better to lead the charge of prominence in manufacturing than those who have served our country? It is a very patriotic challenge we have in front of us to bring manufacturing back to greatness in the U.S. I think that offering some of the young men and women who have been serving our country the opportunity to teach, train and mentor the next generation of manufacturing workforce just might be a match for all parties concerned.

A CONVERSATION WITH BOB WEISHEIT
INDUSTRY MEMBERS – GET TO WORK

Much as Laz says that industry has to be part of the solution to "develop instructors," a good friend and customer, Bob Weisheit, says that industry also has to do more to attract the next generation of manufacturing workers. As long as I have known him, Bob has always taken an active role in mentoring young people. Interestingly enough, most of the leaders in his company are those he once mentored as young people. (Bob will also have more to say in Chapter Thirteen.)

I would say the short and rude, but honest statement: Stop complaining that you can't find people, and get to work. There's more than just being willing to take in a non-experienced person and put a skilled person next to them to patiently teach, train and work with them.

There's more than that. The workplace has to be a place — "Would you want your son working here?" You have to make the workplace interesting, clean, safe, to the extent that you can. I fully understand that a steel mill or a cement factory cannot have as nice of a workplace environment—because of what they're doing and building—that a machine shop can, or that an insurance agency can.

I would tell my peers that the first thing is to make sure you look around and make sure your place is safe, well lit, as good of air quality as you can. A place that if your kid came home from high school and said, "Dad or Mom, I took a job at a machine shop and I want to show you it sometime. I want you to come some day, and I'll give you a tour. It's this great place."

In this chapter, we have brought up many thought-provoking issues. There are so many influencers both good and bad that our youth have in today's culture. We need more industry participants like Bob Weisheit in our workforce stepping up to guide and mentor our next generation. We desperately need them, our guidance counselors, teachers and coaches, to fill this void for the betterment of all of us and our country and not just for their own self gain. Keep in mind that young intern, that mentored a boy or girl, could be America's Next and Greatest Champion. Together we can change perceptions, give our young people so many more choices and expand our skilled workforce and prosperity for our country.

KEY INFLUENCERS ON TEENAGERS AND THEIR POTENTIAL FUTURES IN MANUFACTURING, CONSTRUCTION AND OTHER PATHS UNKNOWN

"We've got a number of larger companies, some employees, some of them engineers, that all have interest in giving back and helping young people learn. Maybe they are retired, and they want to come and put on a class, and they have a special skill that they would like to share with others. That's what it's all about. It's a community kind of an effort, a grassroots effort."

– Gary Skoog, Former President of Golden Corridor
Manufacturing Partnership

There are many entities in this country that have an impact on our young people. We need to encourage each other to serve them better. Companies need to quit thinking about next quarter's results and think more altruistically about our country's future, by way of their future. Adults need to look beyond their own children and consider how to impact others too. Maybe it is a child who is being raised by a single parent, or a child who is reared with little guidance.

Whether it is positive influences like guidance counselors, internship programs, job shadowing, sports teams and activities or negative influences like drugs, gangs and over dependence on technology, we need to encourage and empower the positives and educate and empower against the negatives. We all need to give our young people the best we have. Volunteer, mentor and pay it forward with your time, your compassion, your values, and your experience. Not all young people will want and accept it, but there will be a time when they will need it and embrace it. We all need to be there for them. I encourage you to get on board!

A CONVERSATION WITH VINCE BERTRAM
PLTW AND AUTHOR ON THE EDUCATIONAL ECOSYSTEM

For over a decade, I have been working with high schools in the local communities surrounding suburban Chicago. In these recent times, there have been thankfully a resurgence of shop classes which were once dropped from the curriculums. The main driving force behind the new programs is Project Lead The Way. This program calls for project based learning in a way that combines hands on learning with math, science and other STEM related subjects. I recently called and struck up a conversation with Vince Bertram, President of Project Lead The Way (PLTW) and author of the book called *Dream Differently*. Vince will also talk more on PLTW in Chapter Seven.

I think of the whole ecosystem. What are the things that influence students? You've talked about parents. That's a critical piece that's often missing. Certainly educators, but even narrower than educators in general are guidance counselors, teachers, and principals. There are other people in the community who can influence a student. What's the role of the employer in a community, particularly when you get into things like manufacturing and helping students understand opportunities within their own communities and beyond? I think the more people we can engage in those conversations and take responsibility for those outcomes, the better our students are going to be served.

GUIDANCE COUNSELORS
THE KEY TO PATHS UNKNOWN

There is a need for a rebirth/re-definition of the guidance counselor position. We need to empower these individuals with the reality of what manufacturing careers have to offer. There has been a dramatic change in what the guidance counselor's role has been in schools in the United States over decades. I think it important to describe the evolution of the guidance counselor and the challenges that those in the position have.

In Guidance and School Counseling - A Brief History of School Guidance Counseling in the United States, the article points out *"that the guidance and counseling began in the 1890s and it was concerned with guiding people into the workforce to become productive members of society. These early years were considered to be mostly vocational in nature, but as the profession advanced, other personal concerns became part of the school counselor's agenda".* [1]

As we now near the year 2020, I think it is clearly evident that there is need for a division of responsibilities of counselors. The responsibilities are too broad, the ratios of students to counselor too many and the tasks too significant to be handled by just one defined professional. In order to gain significant traction in the schools, we should have career and behavioral professionals in our high schools, each having a uniquely different focus. This would offer more clearly defined accountability measurements and goals for each area of focus. There would have to be a focus of training on each, especially in the changes in the job and post-secondary educational markets being what they may. The development and hiring of technical instructors is another aspect of the mountain that is needed for us to climb in order to have success at any level of attracting and guiding the "Next Generation" workforce for our research laboratories, engineering departments and production factories of the future.

ISSUES, MAJOR TRENDS AND CONTROVERSIES

In the same article, there are a few major *"issues identified in the guidance counselor profession: what the professional title should be, how counselors should be evaluated, and to what extent counselors should work on prevention instead of remediation."* [1] Discussion on ratio of students to counselor was also described as being an issue needing to be resolved.

With the history mentioned above, it would appear that not only do guidance counselors not have the available time (given additional duties and how many students they are responsible for), many also lack a clear understanding of the ever-changing job market and the desperate needs that have surfaced in the manufacturing and construction industries alone. So many cultural changes have come into play impacting both the students and the guidance counselors and their ability to effectively answer the bell of potential careers—other than a four year degree. As a result, a large percentage of our youth make uninformed decisions, and the potential resultant outcomes are either a failed college experience, no formal education direction at all, or a successful college degree with an uncertain chance of job placement and a mountain of student college loan debt. While many still have a positive collegiate experience, with tuition paid for by either family resources or scholarships, there is still no solution for many of our youth or many industry members in dire need of the same outcome, a skilled energetic workforce paid fairly for an exciting and challenging career.

COMMENTARY FROM VISIONARY LAZ LOPEZ

Having said all of this, there are visionaries in high schools like Dr. Laz Lopez—from the Northern Chicagoland suburbs. (We heard from Dr. Lopez in Chapter Four and will again in later chapters.) Laz also comments on the expanding role of guidance counselors.

It's even supported and recognized by their own national association. It has to include career counseling. The crisis we do have is in the professional development needed and in the experiences we need to provide our counselors for them to knowledgeably represent and talk to students about potential career opportunities.

If you're starting in school now, most of the jobs that you'll be participating in may not have existed. Think of all the jobs that are happening currently that we never even knew or even heard of even three to five years ago in the technology industry. iPads are only out for about seven or eight years. They didn't come out until 2010. Writing codes for apps didn't come out until after 2010. We don't know what those jobs are. The crisis is in how do we help counselors have the tools to help students discover opportunities? Where are they going to fit in the world when that is a space that few of us actually can fully understand?

GUIDANCE COUNSELORS DRIVE THE PATH
BUT HELP IS NEEDED

My suggestion is to even consider a division of responsibilities creating two separate positions given the rise in the needs for various guidance related responsibilities. Manufacturing is making a comeback to the U.S., and automation and technology will lead the way in our ability to compete globally. With these factors, higher skilled positions will become more available and necessary. The only way we have to allow counselors to get career-ready counseling is if industry takes the reigns and helps educate them on the needs that manufacturing, and even construction, have in the marketplace.

Too many times the tools of measurement are not keeping up with the changes in the marketplace. In schools in general, we are getting accountability measurements that push children/students into an outdated model to success. There are disconnects between perceptions and reality.

According to an article by The Education Trust (with support from the MetLife Foundation) "Poised to Lead: How School Counselors can drive College and Career Readiness," the following things should be considered. [2]

. . . When school counselors allow 20% of their students to take up 80% of their time, in effect, 80% of the students do not properly prepare for college and career.

The average counselor-student ratio in our country is now one to 459—Almost double that recommended by the American school counselor association, which is 1 to 250. In some schools and districts the numbers are much higher than the average.

Our country desperately needs school counselors who believe in our underserved youth and are committed to opening the doors to a productive future for all of them.

LAWS TO MAKE CHANGE IN CAREER PATHWAYS

There will be many needed changes to make the impact that will bring the hundreds of thousands of workers needed into the skilled positions that are currently going unfilled. Companies cannot expand and/or book additional business due to a void of workers in their factories. Meanwhile, thousands retire, exiting the workforce in a significantly faster rate than those jobs that are going overseas. It will also take legislators introducing what we need to accomplish these lofty goals and desperate needs. Colorado may be leading the charge with some new innovative laws of their own. Monte Whaley talks to a new law in his 2017 article in The Denver Post. [3]

A law that goes into effect Wednesday mandates public schools in Colorado must inform high school students that not all post-secondary paths lead to college. School counselors also must tell students about jobs as skilled laborers and military personnel.

"A four-year college degree may be a good fit from some," said Phil Covarrubias, the Brighton Republican and owner of an excavation company who sponsored House Bill 1041. "I want students to know that there's great opportunity in trade schools and through military service that doesn't require the enormous cost of tuition at universities."

The bill requires that each student's Individual Career and Academic Plan include information about the various career pathways available to them and the types of certificates and jobs to which each pathway leads.

The law will help reintroduce skilled trades to high school students, who can earn early apprenticeships and exposure to good-paying jobs right after graduation, Covarrubias said. Labor officials say there is a shortage of skilled workers to fill jobs in traditional trades, Covarrubias said.

A 2015 survey by the Bureau of Labor Statistics and the National Association of Home Builders said there are 143,000 vacant construction positions nationwide, and 69% of the members surveyed were experiencing delays in completing projects due to a shortage of workers. Also, a 2015 Deloitte survey of manufacturing executives found eight in ten said the expanding skills gap will affect their ability to keep up with customer demand, and that it took an average of more than three months to recruit skilled laborers.

After a drop from 2009 to 2012, enrollment in career and technical education courses has surged in Colorado with more than 125,000 high schoolers and 20,000 middle schoolers enrolling in 2015. Officials say many are drawn by the prospects of landing a steady, high-paying job right out of high school while avoiding high college debt.

INDUSTRY MAKING A DIFFERENCE!
WE NEED TO LEAD THE WAY!

In Chapter Two, I talked about being interviewed on a Channel 2 TV show. While on the way back in the car from the television interview, I got a call from a Northwestern marketing student who was looking for a story on manufacturing. I thought FOR SURE this young woman had seen the interview. Ironically enough, she had not. Talk about coincidences? It so happens that we were having an event at our office to introduce advanced manufacturing to teachers, parents, workforce board personnel and guidance counselors. I invited her to come and videotape the event. We had a great event. She posted the video to YouTube and coined it "Iverson Manufacturing." Over the years, we held similar events that were mostly students in hope of making a difference one group at a time.

A CONVERSATION WITH GARY SKOOG
FORMER PRESIDENT OF THE GOLDEN CORRIDOR MANUFACTURERS PARTNERSHIP

During that event, we interviewed different adults who were in attendance. One of the adults that day was Gary Skoog. Not only was this video posted on the Medill website, but *Modern Machine Shop* also did an article in the March 2009 edition of their magazine.

Soon after, Gary introduced me to a group that was being formed by manufacturing companies along the 90 West expressway. It was being called "The Golden Corridor." I got involved because there were so many common things that we believed in that would help our young people find their path into manufacturing careers. Gary speaks to the formation of "GCAMP."

I was in economic development at that time working for the Village of Hoffman Estates when I was at an international conference on innovation in North Carolina. I was pumped up by hearing about all of these research parks around the world. I come back and some of my peers think—we can start our own research park. We put together a presentation. We went over to manufacturing companies that are prominent along the I-90 corridor. We laid it out to them and they said, "You know what? We can't find enough good workers. We can't find the number of high skilled workers. If we're selling, our customers say, "Hey, we could buy more if I could find someone to run the equipment." We did the pivot. We said, "Yeah, we hear ya."

Our mission really is to create a sustainable manufacturing workforce along the I-90 corridor, which has been coined The Golden Corridor. We picked up that name and Golden Corridor Advanced Manufacturer. We put Partnership on there because we are partners with municipalities, with the private sector and with local education entities—secondary and postsecondary. We've recently came to a realization that you cannot wait to start talking to high school kids when they are a senior and hope impact their choice on a career path to go into manufacturing or engineering.

THE MAKER MOVEMENT

The Maker movement is something that I hope changes the culture in our country. Wikipedia [4] defines the maker culture to be as follows: *a contemporary culture or subculture representing a technology-based extension of DIY (do it yourself) culture that intersects with hacker culture. Typical interests enjoyed by the maker culture include engineering-oriented pursuits such as electronics, robotics, 3D printing, and the use of Computer Numerical Control (CNC) tools, as well as more traditional activities such as metalworking, woodworking, and, mainly, its predecessor, the traditional arts and crafts.* [4]

We've obtained some state funds for STEM grants and used our own money to fund STEM activities—everything from teaching kids to build and fly drones in fifth grade to a robot called Sphero SPRK. It's the size of a softball, and kids in fifth and sixth grades end up programming it. You download an application on your phone or the school's iPad. You learn how to do some programming in C++ to operate this thing. Kids and parents love it, let alone school administrators.

Another way to entice young people to get into manufacturing and getting into making things is through a makerspace. We are funding a robotics club at Harper Community College. I interviewed some of the kids and took some video of what they were doing. They were building a robot for competition in February. When I asked, "Why are you doing it?" one of the kids says to me, "Well, it's fun. I'm learning. I'm putting my design skills and things I'm learning in school into a project." That's really the bottom line. It's really applying what you've learned in some sort of project and maybe even in a competition from an academic foundation.

We should definitely add a work component. I know some schools are moving in that direction. I know School District 214 here in northeast Illinois, is allowing and encouraging everyone to do an internship by the time they graduate. I've worked in higher education and ran an internship program at a four year college, and I can just see the huge difference when somebody would come back and be able to put it on their resume and build a portfolio from things that they did and knew people and talked to them about the industry. That's how you decide on a career. You don't take some interest inventories or skills inventories. You really just try it. You've got to get in and do it.

How does the educational system train and prepare somebody when things are changing so quickly? The new mantra is, "It's not what you know, but what can you learn?" You're going to be constantly learning and relearning your entire work career because things are changing so quickly. I think we definitely need to look at our education system. Too many of the high schools are evaluated by their state and local school boards on how many of our kids are going to college. What about a two-year college? What about a training program to become a machine operator? Your average manufacturing wage four years out is going to be $50-60,000. Our educational system could use some new blood and some new direction.

I would say the most impactful thing for a young person who is growing up in today's world is having work-based learning that's going to have a significant impact on their decision on which direction they're going to go. There are a lot of different roads and a lot of different bridges to make that happen. The bottom line is impressing upon students, parents and educators that hands-on learning is really a good way to go.

A CONVERSATION WITH MARK HIBNER – PALATINE HIGH SCHOOL

THE MAGIC'S GOING TO HAPPEN

Mark Hibner, as mentioned previously in Chapter Four, is one of the best technology education teachers Illinois has to offer. Here are insights that shed light on the dire need for manufacturing to be a bigger faction of the high school educational experience.

I've been an educator for sixteen years. I became a department chair, actually, in my fourth year before I was tenured as a teacher. The following year, I became the district chair. That was neat, to be able to be in those leadership roles very early in my education career. I helped lead the implementation of the Project Lead The Way Engineering Pathway of Study in our district, with department chairs at the other four high schools. We worked together in building a whole engineering pathway of study with Project Lead The Way. I've also implemented the Advanced Manufacturing Program. We're currently in our fifth year of implementation.

What we've been able to do as we've implemented so many classes, we've changed the face of what the Applied Technology Department looks like. We've introduced so much technology into that lab, helped the persona and perception of what the Applied Technology Department is today.

I have a very diverse classroom. Some students might be coming in having a very hard life. Some might have single parents at home who might be working the night shift. These kids might be working the night shift. They might be working forty hours a week in addition to going to school. It's really exciting to hear back from them, success stories.

In the last five years, I've heard a lot of talk about the need not to go to college. The four year university. I think there needs to be a mindset change on that. The reason why I say that, is because there might be parents and children who are excited and really want that four year college experience. I think instead of just focusing on, "Hey. Don't go to college. Don't go to college." In reality, kids are going to college because they're going to Harper College to get an Associates Degree in Manufacturing.

I think what the mindset needs to be, it needs to be, "Okay. Here you are now at high school. You gained this interest in manufacturing. What is going to be your next step? Your next voyage? What is your next pathway to get you into a manufacturing career? Does that mean you're going to go ahead and enlist in the military, and your military occupational job is going to be a machinist or a welder? Or I need to get hired by a company and have that company put me through Technology & Manufacturing Association (TMA). Or I'm going to pursue an apprenticeship and go through Harper College in a cohort position. Or I am going to pursue a four year degree from Illinois State University in their Engineering Technology program, which is their manufacturing program. Or go to Bradley University, or go to Northern Illinois University. Educators should be responsible for knowing all the opportunities that are available to students post graduation so that they can help students choose the path that is right for them.

Technology Education teachers also need to be able to market the awesome opportunities that exist by taking technology classes. If we want students to be excited about our program, we must sell a quality product and a quality service. It's just like in business. If you're not selling a quality product or service, you're not getting the business. If you're selling a quality educational service to our kids, then those kids go out and spread the word. It's them recruiting for you. It's important that you need to bring all the stakeholders into play. Your counselors, your administrators, and the other educators in other departments. They need to know about manufacturing and how exciting it is and everything that's applied in it. I think that's really important. That involvement is huge. I think I've always believed in that quality product and quality service and also bringing everyone together. If you bring people together, the magic's going to happen.

GANGS, DRUG ABUSE
THE KILLER OF DREAMS, FUTURES AND LIVES

One Rand's internship program is for people with challenges. These people have very real issues. We're talking about living in areas of Chicago that have gang fights. *"One intern says he doesn't need OSHA, he ducks bullets at night. We've actually had an intern that was killed in a gang war the day before he was going start his internship. Hunger is a real issue for these young people. There's no food on the table, they're starving. So that's why the MCIP program is great for those kids."*

Rand is making a statement about some young people who have influences that are beyond description and understanding. Most of us can't comprehend what these young people are going through. Having said this, I have dived into some of the crossroads of gangs, drugs and manufacturing communities.

The MAPI Foundation analyzed the intersection of the current drug crisis and manufacturing to understand the risk it poses to the sector's long term health, and three findings revealed in its report "Ignorance Isn't Bliss. The Impact of Opioids on Manufacturing" stand out.[5]

1. Counties seeing the highest drug-related deaths intersect in an alarming way with manufacturing.

2. The drug crisis is accelerating in communities with large manufacturing workforces.

3. If the trend continues unchecked, the report continues, it will have profound effects on manufacturing in the future.

There are both positive and negative influencers for our youth in this country. So many times employers have told me that they cannot get people who are interviewing to pass a drug test. The interview process is going along great until the drug test is scheduled and, low and behold, that is the last time the person interacts with the employer. Others take a drug test and fail, knowing that it is a requirement.

The Substance Abuse and Mental Health Services Administration (SAMHSA) conducts the annual National Survey on Drug Use and Health (NSDUH), a major source of information on substance use, abuse, and dependence among Americans twelve years and older. Survey respondents report whether they have used specific substances ever in their lives (lifetime), over the past year, and over the past month (also referred to as "current use"). Most analyses focus on past-month use. The following are facts and statistics on substance use in the United States in 2013, the most recent year for NSDUH survey results. Approximately 67,800 people responded to the survey in 2013.[6] Here are some of the highlights:

Illicit drug use in the United States has been increasing. *In 2013, an estimated 24.6 million Americans aged twelve or older—9.4% of the population—had used an illicit drug in the past month. This number is up from 8.3% in 2002. The increase mostly reflects a recent rise in use of marijuana, the most commonly used illicit drug.*

Drug use is highest among people in their late teens and twenties. *In 2013, 22.6% of eighteen to twenty-year-olds reported using an illicit drug in the past month.*

There continues to be a large "treatment gap" in this country. *In 2013, an estimated 22.7 million Americans (8.6%) needed treatment for a problem related to drugs or alcohol, but only about 2.5 million people (0.9%) received treatment at a specialty facility.*

Most people use drugs for the first time when they are teenagers. *There were just over 2.8 million new users of illicit drugs in 2013, or about 7,800 new users per day. Over half (54.1%) were under eighteen years of age.*

More than half of new illicit drug users begin with marijuana. *Next most common are prescription pain relievers, followed by inhalants (which is most common among younger teens).*

While the age-adjusted rate for drug-poisoning deaths involving opioid analgesics has leveled in recent years, the rate for deaths involving heroin has almost tripled since 2010.[7]

With all of these statistics, all of us think, "That would never be in my family!" While at some point of my life I might agree, I can no longer think that way. Everyday I am reminded this is all too close to home.

MIKEY SANTINI
MIKEY'S MESSAGE

The opioid addiction is an epidemic destroying our young men and women, one overdose at a time. Employers are making comments that we cannot find workers who can pass a drug test. I was interviewing for manufacturing positions in my company and had two candidates in a row agree to take drug tests but mysteriously not show up for them nor have any contact with me afterwards. We cannot comprehend the extent of this issue.

Probably the most eye-opening story I could possibly share is that of a young man, Mikey Santini, that my youngest son went to school with in suburban Chicago. They used to run around together on the soccer field when their older sisters played on the same travel soccer team. This young man was an unbelievable athlete. He was a charismatic person with a smile that would melt the coldest of hearts. You could not help but love him.

Somehow this young man got involved in heroin. Try as he may, he could not quit the drug. He could not stay straight. Tragically in 2015 at the age of twenty-five, this young man lost his life to this killer. You could not find a more loving family behind this young man. Mikey had two sisters (Jackie and Jessy), a mom (Geri) and a dad (Bob) who all worshiped the ground he walked on. Everyone loved Mikey. He was just that type of young man. I loved Mikey.

While on a family trip, Cameron, Lindsay and I got the word that Mikey had died. Sitting at a dinner table with family all around—shock and sadness had hit all of our faces. With disbelief we both sat and wondered how can something so tragic could happen to one of our best and brightest young people? It doesn't matter where you live, how affluent or poor your family is, we are all at risk with these dangerous experiments that our kids try. The reasons that they try these things vary, unfortunately some feel the need to lessen their painful, uncomfortable or unpleasant feelings. In a flash, their dreams, futures and lives are cut so tragically short.

I attended this young man's funeral. We all felt a pit in our stomach and our hearts. The funeral home was packed. Soccer players, coaches, teachers, family and friends all lined the funeral home. We all felt their loss. The Lake County (Illinois) sheriff spoke at the service. He reminded us that this epidemic scars so many of our families, and that it has to stop. Sadly, this story is not uncommon, even in affluent suburbia USA. High schools have Naloxone (sold under the name of Narcan), a heroin antidote, now in their hallways. How tragic is this that this problem is so prevalent that we need this in the schools available at a moment's notice—before it is too late!

Back in 2008, a twenty-year-old young man from Buffalo Grove, IL, Alex N. Laliberte died from a heroin overdose. In 2009, Chelsea Laliberte Barnes founded a 501c3 foundation called Live4Lali in order to raise awareness of addiction, prevent substance misuse and its associated harms, and provide access to health care and support services for individuals, families, and communities. In 2015, Chelsea won the Kaleo's Advocate of the year Award. In 2016, the organization won both the Arlington Heights Hearts of Gold and Lake County's Juvenile Officers Association President's Awards.

Mikey's family also started a golf tournament called "Mikey's Message" in order to support the same message for Live4Lali. A close friend of Mikey's also passed from heroin overdose as well. Mikey's sister Jessy has been very involved with working with children with autism and is also an advocate for substance use treatment. In 2017, Jessy was awarded Roosevelt University's prestigious Matthew Freeman Social Justice Award. The entire Santini family has worked tirelessly in their quest to "save one life at a time, and as a result one family at a time." Out of tragedy, the Santini family has looked for what good can be passed forward. It is very apparent the love the family shares for young people and the legacy they will create in Mikey's name.

Once again, we need to offer our young people the best we can offer. Watching them succumb to this deadly disease is scary and so frightening. We need to do whatever we can to break the culture that does little to prevent its young people from ending their lives that had so much promise—lives that are cut short by sophisticated drug cartels and ineffective policies and programs that are not treating this epidemic like the public health crisis that it is.

We need to ALL carry the flag and help find AMERICA's GREATEST CHAMPION. Whether it is a:

- *Young person finding their way into a new manufacturing career,*
- *Young person conquering his or her demons with overcoming drug addiction,*
- *Finding a mentor who volunteers his or her time helping young people, or*
- *A military veteran trying to make their way back into the workforce*

WE ALL must do our best to elevate others to becoming their own CHAMPION Now!® — and as a result make our country a better place.

In our next chapter, we will consider and contemplate the awesome responsibility of parenting. Our culture has introduced so many challenges for our children. We have to work hard at preparing them for success in life.

THE AWESOME RESPONSIBILITY OF PARENTING ALL WITHOUT A BLUEPRINT

"What I've seen with our generation of parents is that too often they try to be friends with their kids. They want that idyllic thing, I see parents going too far out of their way to be their kid's buddy, and that's not quite what the role should be."

– Wayne Larrivee, Radio play-by-play voice of the Green Bay Packers

Parenting is one of the biggest responsibilities we have, a duty we were specifically chosen to perform, a calling. To respect and honor that gift requires hard work and time in understanding who our children are and who they can become. That's not easy because we make decisions based on what we want and our life experiences. That's human nature. The world revolves around our beliefs, desires and goals. In order to nurture the inner Champion of our children, it serves *them* for us to put these things on the shelf. Point: the parenting responsibility and our role as guides to the inner Champion of our children is not something to be taken for granted. It is something to be honored.

No matter your spiritual beliefs, the fact is that children come by way of a most remarkable path. They are miracles. And I believe the honor that God has granted us to bear and raise children is one that only those who have become parents can appreciate. We can all agree we owe them our best. All children's entry into this world is spectacular.

OUR PARENTAL COMPASS
IT NEVER COMES WITH A MANUAL!

There is nothing more important than the parental skills we employ. We should all self-assess to make sure we do the best for our children. This is not always defined or easily done by our current culture. Instead, an objective assessment of each child's talents, skills and personalities is the new go-to strategy.

Our youth deserve the best. They are our future. They deserve the best in education, the best in mentoring, the best in parenting, and the best in coaching. WE ALL MUST DO BETTER. I believe we have an epidemic in this country, and our youth are the ones paying dearly. Our parents generation, and even my generation, learned many hard lessons. We unknowingly all want to make things easier and better for our children, all the while naively short-changing them in the life lessons category. Young people today have it hard. They have much more at risk. Times when my generation and our parents were young were much more forgiving. The stakes were not as harsh as they are now. So by trying to make things easier for our children, sometimes we are postponing their maturity. We are prolonging their childhood, sometimes to the extent of them living in our basements without a job, or a career.

Some of the values that I speak to when I parent, when I coach, when I mentor are:

FAITH: There are clearly many other people more religious than I, however, faith has played a big part in my life. James Preston is our minister who I absolutely think is awesome. Being able to identify and connect with my pastor is vital for me. Kathy and I have a lot to be thankful for. We have three great children and six glorious grandchildren. Life has been good. No, pardon me, it has been GREAT. There have been two particular frightening times that I cannot forget. For the rest of my life, I will forever thank the Lord for the gift of life of our youngest son, and our eldest grandson. In no way does this diminish the other two children, or other five grandchildren and the love we have for all of them.

ACCOUNTABILITY: I witness so many parents thinking that their children did everything the best and they could do no wrong. Even when they did something wrong, they lacked the courage to make their children accountable. Being accountable is what teaches us right from wrong. We sometimes learn hard lessons of what not to do, and as a result, strive to do better—to act better. Having a child who escapes unscathed with never making a mistake is not a badge of honor, especially when it is mom or dad getting him or her off the hook. Somehow we have decided as a culture that accountability has become passé.

WORK ETHIC: We give our children cars, nice ones, expensive ones. Wouldn't it be better to have them earn their own car? They will learn to appreciate it more. They will learn to take care of it. They will look forward to earning enough money to buy a nicer, newer and better car than they first purchased. Each of our kids ended up buying a used car. Each would pay with summer earnings from their summer jobs. No interest of course. We would pay for the insurance if they paid for the car. It was a great decision, and they learned to appreciate things and work hard for what they wanted.

THE HURST WORK ETHIC TAKES THE 25TH PICK AS WELL AS A DR. DEGREE

It is April 27, 2018, and another National Football League draft starts. I am in New York at an event for used a machine tool dealer association, overlooking Central Park. Two high school friends (who married), and their son sit at their home anxiously while the draft starts. He is expected to be picked by an NFL team. My wife, knowing I am interested in the outcome, texts me the results pick by pick on my phone. I walk past Central Park back to my hotel, looking to see if this young man's name is called. Sooner than later, Baltimore picks at number twenty-five. League Commissioner Roger Goodell announces, "The Baltimore Ravens pick Tight End Hayden Hurst." Goosebumps go up and down my arms. This story is the same written by every sports writer in the country. What I want to write is the backstory, not only about this inspiring young man, but his sister and their awesome mom and dad.

Jerry and Cathy Hurst are Hayden's parents. Jerry played football with me in high school. Cathy was one of the first girls I met during my first (7th grade) year at The Bolles School in 1972. It was the first year that they allowed girls into the previous all-boys military school. Cathy was also on the cheerleading squad for the team that both Jerry and I played on. Kylie is their daughter, and there is more to both of her and Hayden's stories.

The reason I choose to include the family in Hayden's story is that this book is about family, mentoring, parenting and the attributes that make people successful. My family story is told in many facets, all the while integrating the manufacturing storyline, but that is not the entire focus of my book. The Hursts have mentored two fine young adults.

Let me tell Kylie's story first. She had a dream to be a vet. Getting into vet school is not the easiest path to success. Kylie was denied entrance to vet school due to lack of large animal experience. Kylie was not dismayed. She went to work for a year at the University of Florida Dairy. Up at 4:30 every morning to begin work at 5:30 a.m., she worked side by side with the staff veterinarians. Before long, she was doing all the calf de-hornings, as well as all the bull calf sterilizations at the dairy. When veterinary student rotations came through the dairy, Kylie was demonstrating to already enrolled veterinary students how to perform those procedures. She was admitted to veterinary school the following fall. Four years later, Kylie added Dr. in front of her name.

The recurring theme in this family is working hard. Both parents Cathy and Jerry's work ethic is exemplary. Jerry has mentored in so many ways, as he works as a dean at a local Jacksonville, Florida high school. As an athlete, Jerry always worked hard. He wasn't the fastest player on the team, but he was the hardest worker. Cathy became a successful professional sales person. She worked tirelessly, while trying to balance time with the family. I can relate, as this has been my chosen profession, but instead in the machine tool world.

Back to the story of Hayden and the Ravens' first pick in the 2018 draft. Let's rewind back to summer of 2011. Cathy and Jerry call me to tell me Hayden has been chosen to participate in the Under Armour All-Star Baseball game. Yes, I totally changed sports. I said baseball. This was Hayden's first and foremost preferred sport. He had extreme success in this sport and was asked to play in the High School All Star game for the entire country. I met the three of them down at the game and took some photos of him playing in my favorite major league ballpark—Wrigley Field—one that I attended games as a young boy, watching my beloved Cubbies.

Almost a year later (2012) Hayden is drafted by the Pittsburgh Pirates. While he was expected as an early round pick, he somehow inexplicably fell to the 17th round. He and his family were disappointed. Nonetheless he got a signing bonus, and off to Bradenton, Florida he went to play for the Pirates Marauders farm team. For three years, he tried to achieve the success he had grown so accustomed to. Hayden was blessed with the frame of a warrior at 6' 5" 220 pounds. He threw the ball ninety plus mph. He had it all going on. But something just didn't click in the minor leagues. Try as Hayden might, he could not achieve the same level of success that he had at the previous level. Doubt set in first. He then became distraught. He turned to his family after efforts with his coaches were unsuccessful. There was a lot of discussion and ultimately Hayden decided that with the support of his family, that baseball was done.

It was about this time that Jerry and Cathy had lunch with Kathy and I. True to form, the parents invited their children to join us, and they did! This is not always the case with young adult children. Just a small example of what great young people both Hayden and Kylie are. At lunch, Hayden tells us that in fact his baseball career is over. He tells of his plans to try out at the University of South Carolina, where a childhood friend of his is playing quarterback. He had spoken to Assistant Coach Steve Spurrier Jr., and they were both confident that he could compete in the SEC, one of the highest levels of collegiate football. Hayden goes on to say that he last played at Bolles in his junior year, and that his senior year was totally understandably focused on baseball.

Hayden went out and took Columbia, SC by storm. He out-lifted almost everyone. He out-sprinted almost everyone. He took what God blessed him with and formed himself into a NFL-worthy player. This is the backstory that speaks volumes to Hayden and the rest of the family. Like Kylie, he took the long route. No shortcuts for either.

I have hired dozens of young people who want to establish their careers as professional sales people. So many want to take shortcuts. Every time I explain that you have to understand the right way to do something—yes the long way. Until you have mastered that, only then can you understand what the shortcut even looks like. Both of these young people get this, of course because of the priority their parents have put on it.

Fast forward again to press conference day in Baltimore at Raven's Headquarters. During all the introductions and all the interviews, one thing cannot be missed—Hayden's "Yes Sir's and No Sir's." Each and every one features the same respect-filled commentary. I ask everyone to take note of several aspects of the Hurst story. Whether it is to never give up and persevere, work the hardest in any organization or team, or be respectful at all times, all of these lessons are spot on. I cannot be happier for these two fine young adults and their paths to greatness.

RELATIONSHIPS: Today's children have a tough time maintaining and developing relationships. When we were children, we did not have the internet, email, cell phones and texting. We had to learn to describe what we wanted, communicating with others and developing friendships. We made bonds, many times, that last a lifetime. In today's culture, that is not nearly as easy as it was when we were young. Me on the other hand, I go overboard. As I mentioned, moving sixteen times during my childhood certainly allowed me to perfect the art of meeting new people. Each move became an opportunity to make new friends! I only hope that the apple has not fallen too far from the tree, and our kids perfect the art of relationships—despite technology pulling their generation into the relationship abyss. Life is too short and meeting people and learning their story is awesome.

HONESTY: This has always been a black and white issue for me. White lies didn't make sense. Justifying such just meant to me that telling bigger lies are just around the corner. When the stakes rise, there is more on the line. We always taught our children to tell the truth. I told my soccer players the same. Lying did not fly with me. My mom taught me that if I couldn't tell her about what I was doing, chances are I shouldn't be doing it. Of course my mom made it easy for me to tell her anything.

ETHICS: Each year I see the stakes get higher. The rules and lines that are drawn on behavior become fuzzier. What is and is not acceptable is now up to interpretation. When we were young—right was right and wrong was wrong. The WHY did not matter. It didn't matter who was playing, or who his or her parents were. Little things make a difference. Ethics are something that survive the test of time. Ethical behavior is not always cool, not popular and often will not get you brownie points from many. However what you choose to do when NO ONE is watching defines you. That feeling cannot be bought; it cannot be given. Only you can give it to yourself. I am thankful for my ethical and moral compass. Though I am certainly not perfect, I thank my parents for giving me so many of these. I try to make them proud—everyday.

INTEGRITY: Tells all about who you are dealing with. I am not trying to say that our children and next generation are not good and genuine. I am of the belief that ALL children are good and precious. I am of the belief, however, that is up to us: parents, influencers, coaches, elders, and educators, to give them our best.

Too often, some coaches are in it for the wrong reasons. They are trying to relive their childhood. They are coaching their children so that they can get more playing time. They are in it for a WIN only attitude. While winning is certainly important, and I am against participant awards just for showing up, it should not overshadow the real reason our children play on teams. We want them to learn life lessons. We want them to learn to work on a team and be a part of something bigger than themselves. We want them to look up to our coaches and learn the honest way to win.

Cheating is one the absolute worst things we can teach our youth. I once watched two little league baseball coaches, at the time of the draft trade, pick numbers under the table because it suited each of their teams better. I told the league administrators that if this was allowed, I would be gone the next year.

On September 11, 2001, we were all mourning as the entire country watched the biggest single disaster in U.S. History (on U.S. soil) unfold on television in New York City. I was sick to my stomach watching the terrorist attack on television in disbelief. It was a day that none of us will ever forget. In our parents' generation, it was JFK being shot, or when Japan bombed Hawaii. For our generation it was 9-11. The rest of the story is the epitome of a bad coach doing the wrong thing to teach his young players and their parents.

The afternoon of September 11, 2001, my U12 soccer team had a travel soccer game about a forty-five minute drive each way. I called the other coach, and I said, "Coach there is no soccer going to be played today. There is just too much sadness and terror in New York." He agreed to reschedule.

Later in the fall, we did reschedule the game—earlier, because it would be dark otherwise. The team was a good match for our team. They played hard, as did we. They were talented, as our team was as well. We started the game early. A second referee showed up and was sent home. The official was running five to seven minutes long on time, blowing the whistle only after the other team scored. The second half had a repeat of an extra long half with a whistle after the opponents' goal. I have never seen such blatant poor sportsmanship and integrity. The bottom line is that the second official was the REAL referee whom we had sent home. Their assistant coach had taken the game to officiate. It was all getting clear to me at this point. The minute I got home, I submitted a protest. (The first and last one I have ever filed.) Not only did we win the protest, that team was not allowed to play the subsequent following season. This is not the type of coaching behavior that our sons and daughters need to emulate.

RESPECT: It goes both ways!

This is a big one that combines many of the other subjects together. My children had to earn respect; they understood that. My players had to give respect to our opponent, to their elders, and to the officials. One young man who played for my team had the misfortune to decide to spit in the direction of a referee. By my reaction, I am sure I startled my player. He was told in no uncertain terms that he would never play for me if he ever did that again. This young man from that point on never crossed the line. This young man became a very respectful stand-up young adult.

Too many times, I saw other coaches condone cursing at adults, other players, even the officials. Too few times did I see any consequences result with their players. I had opposing players tell me off or tell my parents off. It was very upsetting and disappointing. Why a coach or parent thinks this is acceptable, I will never understand.

I ask all of us coaches to RISE UP. Take the high road. Be willing to lose for the greater good. The lessons will last a lifetime, not just one week or one season and be forgotten about. I had to suspend players for various poor choices in judgement. The biggest card I had to play was them letting their teammates down. That moved mountains in my world when I was a kid, and it did the same for my players. They respected me, but more importantly they didn't want to let their teammates down. Team sports are a powerful influencer and definitely improved my makeup/character as a young person.

FRIENDSHIP: PARENTING IS NOT FOR SISSIES
- AND OUR CHILDREN ARE NOT OUR BFF'S

I ask us parents to do the hardest job of all. Don't try to be a friend to your child. That should not be your role. Too many parents are working to support their families. Many times both parents are working. There is not adequate time to spend together. Certainly not as much as any of us would like. As a result, there is guilt. Some parents make decisions based on guilt. Instead of making the right decision that they know is hard—but right—they make a decision that merely allows them to feel better. This is a slippery slope. I firmly believe that once we can acknowledge the feelings of guilt, the awesome responsibility of parenting becomes much clearer and much more rewarding.

PIVOT POINT: Recognize when they present themselves! The question as a parent—would you be willing to go to such extremes to protect your child? Many times we do not get a second chance. Many times we decide that this would get in the way of staying in the good favor of our child. That is my point. You have to make a choice, and it will not always be the correct one. Staying friends with your child should not be a criteria point. Allowing guilt to justify your decision should not either. Protecting him or her and using your experience and best judgement whatever the decision—should. PS: We often explain to our oldest that he was our guinea pig. We tried everything first with him, learning one decision at a time.

MENTORSHIP: Pay it forward EVERY time you can

I ask mentors to be generous with your time. Chances are someone was patient with you at some point in your life. I ask relatives and counselors to influence our youth by exposing them to as many things as you can. So many of us bring our and only our view on life and opportunities without broadening our experiences for their sake. What is right for us, may very well not be right for them.

REMEMBER they deserve our BEST. They are AMERICA'S NEXT GREATEST CHAMPION —with our help!

ENCOURAGE YOUR CHILD TO RUN WHERE THEY ARE DESTINED TO BE

Bigger thought: push your children—not where you want them to go, but instead where they are destined to go. Lead them to the place where they are accepted, cherished, where they are loved. We shouldn't prejudge, or get in the way. Potentially that means manufacturing jobs too. The culture is going to push them where they think they should go. In our culture, people will say, "You need to do this or you need to do that" or "You're a 4.0 and good at math and science so you should be a doctor." How does that young person respond? He/she oftentimes says, "Yes, but I don't want to do that." They may be tempted to follow culture down the wrong road. Why? It's scary for us to go off script. Instead, tell your child, "Just don't run in place. Run where you want to run. But don't walk aimlessly either. Don't stand in place." They still may ask where. When they do, you already have the golden answer inside you. It is simple: go figure it out. We also need to recognize our child's talents and more importantly their weaknesses. In life we need to get really good at leveraging our strengths and compensating for (or strengthening) our weaknesses. Our children cannot always recognize the difference between the two.

My dad always encouraged me to make my own decisions. He would help me understand and know all the facts before making a decision, then own it. Live up to the commitment and take the credit or the blame—either way. I, myself, sat in the hot seat as a parent. I myself was tested as a parent to listen first and not influence based on my own biases and life experiences.

My eldest son, Britton, was a great student. He took all sorts of AP courses, to the point of getting burnt out. His sophomore year he took three AP courses and was concerned about being able to balance being an athlete with the demands of taking three AP courses. He called me at work one day, asking if he could have lunch with me. I said sure, let's meet up the road about halfway between my office and our house. I really wasn't sure what he wanted to talk about, but it must be important for him to want to sit and talk over lunch. After sitting down and extending some idle chat about school, soccer etc, I asked him what was up.

"Dad, I don't know if I can play soccer and take three AP courses at the same time. I want to ask you what would you do?"

I thought carefully before I responded. Brit, you are going to have to make this decision on your own. Think about it carefully. Make sure you know that whatever you decide, you need to own and contend with the consequences and/or results. Brit while you are my son and we are similar, we are still different people.

Ultimately, Britton decided to quit soccer and focus on his studies. He would later decide to pick back up soccer his junior year. While most of his peers were on the varsity team, he was back on the JV team. The rest of the story is that he ended being MVP for the JV team. His senior year, he did play varsity. In my mind, I would have never given up a sport in order to study for AP classes. Brit was smarter. He had so much pure intellect that he made the right decision for him. If I had told him what I would do in his shoes, I would have said just the opposite of his decision. But my son wasn't me. Fortunately, I had the foresight not to burden him with what I would have done.

A CONVERSATION WITH BUZ HOFFMAN
PRESIDENT OF LAKEWOOD HOMES AND PARENT OF ONE OF MY SOCCER PLAYERS

Buz Hoffman and I met when I was coaching his son in soccer. Buz and his wife Joey are very much from the old school, like Kathy and I, when it comes to raising children. Rarely are we ever on opposite sides of a topic. I will always have the ultimate respect for this couple. Buz owns Lakewood Homes, which was a builder of entire communities in the Chicagoland market. I asked Buz for some of his views on parenting, where the next generation appears to be going, and his view of someone helping them get there.

I can't speak to factory manufacturing, but I can testify to the lack of field manufacturing—skilled tradesmen and the mechanicals. I guess what continues to become closer to my heart is how kids (millennials and such) have been and continue to be raised—at home, at school, by peers, by politicians etc. Being raised that everyone ALWAYS wins. They have no clue how to confront adversity or loss. I'm more focused on what I'm angry at with our new culture than the lack of kids' shunning of careers in manufacturing or in the trades.

This idea that everybody's a winner. Everybody's not a winner. If you don't learn that at an early enough age, by the time you're getting to understand things. You got to learn that you're not always gonna win. You've got to be taught how to deal with it. If your parents, and your "mentors" are not helping you through this process, you don't learn.

Buz's analogy with both manufacturing and building are in the same category and share a great deal of the same skill set. Many people start out in carpentry and end up in metalworking (as did my dad's uncle Stanley). One of my service technicians changed careers and went into building/carpentry. He did the finish work on the house Kathy and I built for our three children to grow up in.

Many kids today are not being mentored by the proper people as to what to do when they are of work age. All they know is what their parents told them: you have to get through high school. Certainly from the area that you and I are from, you have to go to college. Then you go to work.

Frankly, we are producing a workforce that is the most ill-prepared workforce in the history of this country. I get frustrated just thinking about it because they think they have to go to college. They get out, they haven't been trained to do anything "practical." They think some nice cushy job is going to fall into their laps, and they're going to make $75,000 or $100,000 dollars, and go on happily ever after. Everybody is NOT going to graduate and make $100,000.00. They feel that going into manufacturing, going into trade work, is beneath them.

There should be a mandatory class in high school, or series of classes, on careers. This class (or classes) on careers will ask, "What is our economy today? What are the jobs?" It's gotta talk about, number one, first and foremost, "What is the compensation you can expect?"

If you don't show these kids that they're going to make a good living, they're going to go into their general studies in college. You have to show them that they're going to make a good living. "Now that we've shown you what the expected compensations are, then what's the training required?" I can do this one in one year. I can do this one in two years. I can do this in one to five years. What goes along with the training? What are the opportunities past that? If I go into manufacturing or the trades, what are unions? What are union dues? What do I have to pay, and what do I get for the unions? The things that the economy is based on today. They don't have a clue.

What happens if we keep going the way we are? These kids get out of high school. They go to college. They get their degree in art history, and they can't find a job. So they start flipping burgers and whining that they're not getting minimum wage after they've spent four years at college. They can't begin to pay back their student loan.

What's the matter with getting your hands dirty? Why do you think that's beneath you? You can be a union carpenter, and with benefits, you can be at $80,000, $90,000 dollars a year. You don't have the same stress level. You don't have to put on a tie every day. It's almost a sea of change for these kids today. They suddenly deem blue-collar work to be beneath them.

Parents of kids, in so many ways, follow along what the schools are teaching. It has to be school administrations, which we charge the children with getting our kids ready. The majority of this culture change has to come from them. It's got to start really soon. We're running out of workers. There are not enough workers in the northern United States to build the houses that should be being built right now.

While Buz Hoffman does not reside in the manufacturing sector like me, he does speak to the fact that the construction industry has the same exact workforce problem.

Below is additional data to support the case that Buz makes to the lack of skill and adequate numbers of skilled positions for the current market. The one thing that manufacturing differentiates from construction is the influence of technology and innovation. While it takes a great deal of skill to be a carpenter, a plumber, an electrician, a roofer, and so on... the machining and manufacturing arena is employing technology and automation in a much more significant rate. This will require the skill set to increase as the implementation of these items increases. From the Associated General Contractor's 2018 Construction Hiring and Business Outlook Survey:[1]

- *75% of contractors want to increase their headcount in 2018.*
- *78% of firms report that they are having a hard time finding qualified workers to hire*
- *Even as firms expand headcount, an overwhelming majority - 82% expect it will become harder or remain difficult to recruit and hire qualified workers in 2018*

WAYNE LARRIVEE
CALLING THE PLAYS FOR THE YOUTH OF OUR COUNTRY

When our oldest son was young, not only did he play soccer, he also played baseball. During the time he did play baseball, I was fortunate to meet the Larrivee family. Scott Larrivee is Wayne and Julie Larrivee's son. Our son Brit is the same age as Scott. Both my son and Wayne's were on our team.

At the time, Wayne lived in the Chicago Northwest suburbs and announced for the Chicago Bears. He was/is very talented doing play by play then and now for the Green Bay Packers. Wayne and Julie always impressed me in the type of parents they were with their two sons. I asked Wayne to speak to some of the optics that are seen in parenting in our culture and his thoughts on what it all means and the influence it has on our young adults.

You need a special skill, and it's not just something. You can go to the trades on many of these jobs. A lot of it now, with the technical aspect we're talking about here—the computer training—you really need to be in a technical school. You need to be focused on that type of thing to have a chance. If you were a plumber or you got into the plumbing trade, you came up through the apprenticeship. When you're talking about the high tech computer stuff, that's a whole different ball game.

I think that people are so focused on "let's be millionaires" that they lose sight of the fact that the middle class has always been good. There's no question that you can make a great living and raise a family in an upper-middle class type environment with making a living through manufacturing and engineering fields. The better life may not be found going through a four-year school. A lot of them were there, and they were killing time. What I saw later on with my son through college is that a lot of kids weren't quite sure what they wanted to do. I have always felt that if you go to a four-year college, and you don't have a specific idea—a goal of what you want to do, where you want to end up—then you're kind of wasting time and money.

We're starting to see that now, where people are saying there are other things to do here. You don't have to go to a four-year college. Maybe you're not a great student. You don't like school. Well that's fine. That doesn't mean you're going to be a failure. That doesn't mean you're on food stamps the rest of your life. I think that was a great fear.

What I've seen with our generation of parents is that too often they try to be friends with their kids. They want that idyllic thing. I see parents going too far out of their way to be their kid's buddy. That's not quite what the role should be. There is a lot of divorce today. A lot of parents know their kids have gone through a rough patch because of this breakup of the marriage. They're trying to make up for that. All these intentions are good, but by the same token not really doing the kid a lot of good.

I used to do Big Ten games on television. I was working with Randy Wright, the quarterback of Wisconsin and later Green Bay and Pittsburgh, and we went into Northwestern. Even athletically, Northwestern, gets mostly kids who have been in upper echelon type of schools to have the grades to get into Northwestern. They're talking about suburban kids, they're talking about rich kids. Those are the kids that usually get to Northwestern and play football.

The late Randy Walker was the head coach. We sat down one Friday afternoon in a production meeting. He was always great. I guess he had a rough week in practice with the kids. He said "You know what? I yelled at so-and-so the other day, and I think it's the first time anyone has ever raised their voice in this twenty-one-year-old's life." Think about that. He was not wrong. He was probably absolutely correct. Think about the way people raise their kids in affluence: their need to either, make up for the divorce or make up for time missed because you're spending too much time on the road in business, or just trying to be friends with your kid because you think it's a neat thing to do. He says, "We battle this every day. We have to teach these kids. These kids have never been yelled at. They've never had discipline. They've never been told constructively they're doing something wrong. It's always been positive."

There's a couple of small town school districts up here, north of Milwaukee. Every kid must be on that honor roll every semester. What good is that doing anybody? If the teacher doesn't give the kid a good enough grade, the parent goes in there and beats up the teacher, verbally. That's where we are. That's not good. That's really hurting our kids more than it's helping.

I see coaches now doing a much better job of mentoring some of these kids. Just because somebody is divorced doesn't mean their life, or their kid's life is going to be ruined. There are some voids there that sometimes needed to be filled and were filled by mentors.

DIVORCE: LAND OF THE SINGLE PARENT
DON'T GET LOST

When I was a young person, I grew up in a home with one parent—my mom.

My parents divorced when my sister Kelly and I were very young. It was for the best. They both were very strong-willed. Both wonderful talented loving people.

Marriage and parenting is a series of compromises. The reason that I bring this up is that I believe we have a parenting epidemic in the U.S. For a start, only 50% of marriages last. Families are forced to be working on "less than full staff" to coin a familiar term to those of us running companies.

While we all are trying our best, rarely do we have the opportunity to excel. Single mothers (or fathers) are doing the best they can to be the breadwinner, while at the same time function as both the mother and father. Children are, at the same time, learning a great deal on their own. Occasionally, some of us benefit with great mentors while others struggle to learn the finer points of growing up.

Many single parents are ridden with guilt, doing the best they can with the time and energy they have. This transpires into more complicated emotions of giving their children too much, or trying to be their best friend, or not holding their child accountable. ALL of these come out of the same emotion—guilt by the parent for feeling that they are not doing the job their parents did. Of course we all know that those were different times. The world was much more simplistic. Things were more cut and dry, and everything did not move so gosh-darn fast. And much more of the time, there was a double parent functioning family.

I was given the opportunity to be raised by a wonderfully fun and energetic mother. She tried her best and worked very hard. I never felt like I was losing out. I always knew that my mom would give her last dollar to me, even when I knew that she needed it more than I did. Unfortunately, alcohol got in her way, and we lost her to throat cancer in 2002. I am forever grateful to my sister Kelly for her dedication to take Mom to radiation treatments during her fight. In case I never got the chance to say it enough. THANK YOU MOM. YOU DONE GOOD. I LOVE YOU.

Dad always encouraged me to strive for excellence and perfection. Working for him was hard, but he was so talented that I cherished the ability to have been mentored by him early in my career. Initially, I was offered a chance to work for my uncle. He and I have always had a close relationship. We were closer in age.

He was the youngest of the brothers, me one of the oldest offspring. He had all girls, and I think the fact that I was the oldest of the boys was a connection for us both. I did go to work for him right after Kathy and I got married.

When I had been buying material for my uncle's company for about five months, my dad called me and asked if I wanted to sell machine tools for him. My mind raced to answer him in milliseconds—no, but I halted from blurting something out that quickly. I did owe it to my wife, him, and even my grandfather to give this some careful thought prior to giving an answer. I decided to go home and talk it over and then give him an answer the following day. The following day I said yes, and the rest of my career is history, as they say.

I owe my ability to sell, to see what the customers want and make certain that I can exceed their expectations, to my dad. One of the simplest things he taught me when a customer would say or ask something that was totally unreasonable – "Simply ask the question do you REALLY think that is fair?" I would even repeat back to the customer what he was asking of me. In most cases, the customer would acknowledge how unreasonable his or her request was.

DOING THE BEST FOR YOUR CHILD
WHAT DOES THAT MEAN?

As we all look at doing the best for our children, that definition varies from year to year doesn't it? When they are born, we are trying to feed them, change their diapers, keep them from falling, and make sure that they are loved. When they are five, we are making sure that they are curious, that they learn, that they learn how to fit in their environment, and again, that they know that they are loved. When they are ten we want them to have started establishing their identities, so they know they are special and unique, and that they are loved. When they hit fifteen, we try to make sure that they know that the decisions they make on their own have significant consequences—good and bad. They have to find their way in the world and treat people with respect along with earning respect. They start to figure out what they want to do with their lives, and again, know that they are loved. So when it comes to the next big step in their lives, we need to have the conversation about what intrigues them, what talents they feel they can employ in a career, what lifestyle they want and will have to afford, and what the market will bear for their choice. It seems so simple, but yet it is rarely done effectively. Most parents (who have the means financially) think supporting their children simply means to supply them the financial resources for them to go off to college to figure out what they want to do. In actuality, this might be the worst thing for some of them.

While I am not advocating anything is a one-size-fits-all solution, the go to college and figure it out certainly is an overused tactic. Each of our children is different. Doing the same for all is not necessarily fair to all. Supporting your individual children for the individuals that they are may not look all the same. One may want to learn a trade, the other go to law school, while the other may want to be an entrepreneur and start a product or a company at a young age.

In the next chapter, we will investigate the ROE (Return on Education, instead of Return on Investment)—for both those who do not have the money (and have to take out loans), and even those who DO have the money. That same money that would go into a four year education, can have multiples uses: 1) tuition and fees at a two year program (10% of total) 2) finance a purchase of a dwelling, whatever that would look like for him or her 3) put into a safe investment portfolio to earn income 4) become seed money to start his or her own business. Reading on can inspire you to decide what is best for your Champion starting their new lives—where they will be responsible for their own financial livelihood, maybe raise a family and make you and your spouse proud in any number of ways!

ROE: RETURN ON EDUCATION LOOKING AT YOUR EDUCATION PATH AS AN INVESTMENT

"Right now no one is really accountable for student outcomes. It's graduation rates."

– Vince Bertram, President of Project Lead The Way

In business, we look at ROI, the return on the investment we put in. This is a basic concept. If I put money into an investment, I am looking to get more out of it than what I put in. It is basic, simple and practical. Do parents and students look at their ROE? Return on their Education? It is about more than money. It is that but also time and your futures. Return on Education is a bigger picture than just a financial decision with financial rewards. It takes into account a match of what you have studied (with your time and money) and what you will get back in money, happiness and satisfaction in the career in which you have become invested.

In this chapter, I am not trying to convince anyone NOT to go to college. Rather, I will try to open every reader's eyes to some compelling facts about sending a young person to college and how to assess the right choice for them. There are many factors, and each and every person's situation is different. The fallacy is that key factors in deciding what to do after high school have changed dramatically over the past few decades: the job market, the educational model, our culture, and the affordability and effectiveness of college for many students.

MORE OPTIONS, MORE SUCCESS

Parents today are in a quandary, and they have a choice to make. They can ill-afford to think that throwing good money into a bad premise is going to work. A small percentage of our children will succeed, regardless. A larger percentage need the right direction from their parents and the open-mindedness to explore all the possible avenues.

How many students can afford the college of their choice? How many parents put a second mortgage on the house in order to pay for their child(ren)'s secondary education? How many students cannot get accepted into the very competitive college acceptance process at the college of their choice? How many students are taking out student loans that put them in a hole that they cannot get out of?

As we continue in these uncertain times, there are many financial reasons to investigate alternative educations that could lead to very productive and financially rewarding careers. Many technical and community colleges offer two-year degrees that cost a fraction of many four year university degrees. As a result, the student can enter into the workforce making $50,000-$60,000 a year and have a very short payback on their educational costs. The four-year degree student is finding him or herself in debt after having paid $160,000-$200,000 (or more) for their education. They will have a decade or more of trying to pay back these costs. They also have a difficult time trying to find a job and differentiating themselves from others with the same degree. With unemployment now around the 4% rate, maybe making $40,000 a year with experienced individuals willing to work for less, is creating a reverse competitive market for the employer.

The irony: manufacturing companies many times have to bring in engineers from India and other countries because of the lack of supply. Some manufacturers cannot expand their business due to the lack of talent. There is a significant demand for personnel.

MORE OPTIONS, MORE OPPORTUNITIES

Without question, there are jobs available for more money with an education that takes less time and costs less to get. So let's ask the obvious question, why aren't people flocking to manufacturing careers?

Perception and cultural image is the popular answer. The market used to be that "professional positions" were 60% of the requirement in the working world thirty years ago. As a result, an entire generation was told that was the target to educate and prepare students for. Meanwhile, times have changed, and as a result, the market now demands 60% of the workforce to be a skilled workforce. What is lagging behind is the perception by the market. Young people are struggling to figure out what to study and what career choices they need to make.

Author Charles Sykes brings out a number of good questions and challenges us all to rethink the typical education advice given in his book *Fail U: The False Promise of Higher Education*. Here are some of the highlights Sykes points out:[1]

- *The cost of a college degree has increased by 1125% since 1978*
- *(four times the rate of inflation.)*
- *Total student debt has surpassed 1.3 trillion.*
- *Nearly 2/3 of all college students must borrow to study.*
- *The average student graduates with more than $30,000 of debt.*

What, then, can we learn from these facts? Sykes describes a vision of higher education as being one that is affordable, more productive, and better suited to meet the needs of of a diverse range of students and that will actually be useful in their future careers and lives."

Sykes points to private education as the equivalent of buying a BMW every year and driving it off of a cliff. He cites many private colleges like Duke University, Dartmouth College, Wesleyan University, Boston College, and Southern Methodist which command more than $60,000 a year.

Student loan debts exceed both the nation's total credit card and auto loan debt. The delinquency rate in student loans is higher than the delinquency rate on credit cards, auto loans, and home mortgages.[2]

- *A survey of 30,000 alumni by the Gallup-Purdue index found that only 38% of recent college graduates strongly agree that their degree was worth the cost. Only a third of graduates with student debt think their education was worth the price tag. Their skepticism is understandable.*
- *A study by the Federal Reserve Bank of New York in 2012 found roughly 44% of recent college graduates working in jobs that did not require degrees-the majority of them in low-wage jobs*
- *A study by Rich Richard Vetter* and Christopher Denhard found that there are more college graduates working in retail jobs than there are soldiers in the U.S. Army and more janitors with bachelor's degrees than chemists.*
- *Only 34% of students entering four-year institutions earn a bachelor's degree in four years.*
- *Barely 2/3 or 64%-finish within six years*

THE CHOICE

I witnessed the dubiously ineffective educational model. From 1977 to 1983, I attended five different universities. In each case I decided that the educational model, at least from my youth, was ineffective. I found the methodology and learning to be extremely boring. I had trouble keeping focus in the classroom. The terrible lack of challenge and effectiveness was not consistent with the manner in which I learned.

One late night while sitting in my thermal dynamics class at Marquette, I had an epiphany. Listening to the teacher lecture, I had the realization that Kathy and our two young children were at home alone again without me—why? It was very late at night. Several times a week I would arrive home after midnight. Why was I doing that to our family? At the time, I had a full-time job. I had more engineering background and education than my job would ultimately require. I was selling machine tools. I totally understood and was fascinated with making parts, not designing the machine tool itself. In retrospect, I would probably have been best served learning to program a CNC machine tool—but that technology was just emerging in the industry.

I realized that my young family was suffering by my absence more than the benefit (economic or otherwise) of furthering my education. I decided, in that instant, to go home and quit college. I also came to the realization that maybe I was more concerned with what people thought, than about the impact the education would have on my life. The only person whose opinion I cared about was Kathy's. She agreed it was up to me whether I finished school or not. That was the last college class I took. The Bolles School, my prep high school in Jacksonville, was what set me up for success more than any college I attended.

ANNUAL INCOME/AGE

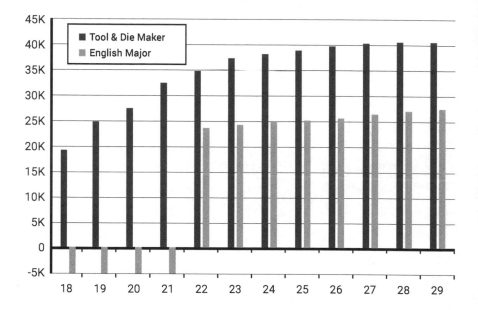

TOOL & DIE OR ENGLISH MAJOR?
A MILLION DOLLAR DECISION

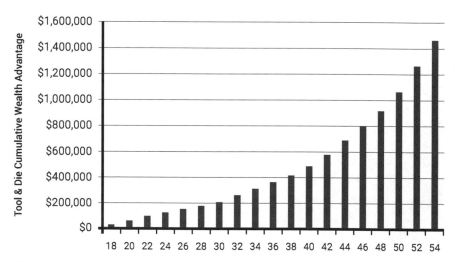

Assumptions:
1. Zero wage inflation.
2. Savings=50% of difference in income.
3. Investment return of 7% per year savings

Source: Reshoring Initiative

CONSIDER THE NUMBERS
THEY DON'T LIE (SHOCKING REVELATIONS)

The reality of the situation right now in our country is that we have millions of young students who come out of school with a mountain of debt from student loans. The website Comet estimates $1.419 trillion in student debt. Just to explain the gravity of this number, let me share some of the statistics from this site.[3]

Americans now owe more than $1.3 trillion in student loan debt, based on the most current figures available to Comet. That money is not only owed by young people fresh out of college, but also by borrowers who have been out of school for a decade or more. The standard repayment timetable for federal loans is 10 years, but research suggests it actually takes four-year degree holders an average of 19.7 years to pay off their loans.

Top statistics of the student loan debt landscape in 2018:
- *Current U.S. student loan debt = est. $1.4 trillion*
- *1 in 4 Americans have student loan debt: An estimated 44 million people*
- *Average student loan debt amount = $37,172 (* see graphic)*
- *Average student loan payment = $393/month*

In the past decade, total U.S. student loan debt has surpassed credit card debt and auto loan debt. In the third quarter of 2017, Americans owed $810 billion on their credit cards and $1.21 trillion in auto loans. Currently, U.S. student loan obligations are larger than both, trailing only mortgages in scope and impact.

Student loan debt has ballooned in the past few decades, primarily because the costs associated with higher education—tuition, fees, housing, and books—have grown much faster than family incomes. The College Board has tracked costs at public and private universities since 1971.

When the organization first started monitoring prices, the average cost of one year at a public university was $1,410 ($8,450 in 2017 dollars). That was 15.6% of the median household income of $9,027 and manageable for many families without *going into debt.*

Fast forward to 2018, and the picture is very different. Today, the average cost of one year at a public university is $20,770, which is 35.2% of the median household income of $59,039. That could be why more than 70% of bachelor's degree recipients emerge from college today with substantial student loan debt, and why many find themselves in need of loan consolidation and refinancing.

There is a lesson here. Maybe we need to stop and think about the ROE on a college education: the time it takes, the costs and how soon a young person can be in the working world making money rather than spending it on the education. Starting out a career with a boulder of seemingly insurmountable financial responsibility on their shoulders is putting a huge strain on our young adults futures.

A HIGH SCHOOL VISIONARY WHO TOOK A STAND
A CONVERSATION WITH LAZ LOPEZ
HIGH SCHOOLS PREPARING YOUTH FOR CAREER READINESS

We have heard from Laz Lopez in previous chapters. Here he speaks to how high schools need to do a better job of meeting the needs of the students but also of the surrounding employers. His innovative vision and development at Wheeling High School has gained him respect and admiration from parents, administrators and industry members alike. Additionally Wheeling High School was one of the very first PRIME High Schools as designated by SME. In order to understand Dr. Lopez commentary let me first explain the PRIME initiative.

PRIME INITIATIVE
PARTNERSHIP RESPONSE IN MANUFACTURING EDUCATION

In 2011 the SME Education Foundation led by the efforts of my friend

Rodney Grover, developed the PRIME initiative, which stands for Partnership Response In Manufacturing Education. The PRIME initiative engages manufacturing companies in the building of high school manufacturing technology programs in their own communities. PRIME provides a platform for manufacturers to charitably contribute to their own community, while fostering their future workforce. The program now in twenty six states, gives high school students the opportunity to gain the skills necessary to immediately find employment, provides financial support to the school for equipment and curriculum updates as well as summer camp programming for middle school students in the district. In addition the SME Education Foundation provides access to scholarship funds for graduating PRIME school students.

DR. LAZ LOPEZ
BEING RELEVANT IS THE KEY

As a school principal, my goal is to ensure that we are being relevant. The value of a high school diploma is widely questioned. One of the strategies the high school leader can use to validate the quality of the education a student is receiving is through the authentic and relevant experiences students can have that give the community insight into the school and into the value it's offering.

If you have an increased number of students being employed in the community as a result of the work that's happening in the high school—as well as those earning professional degrees or credentials in demand and returning to the community to work—that is a direct validation of the quality of the school.

A high school can serve as a lever for broader economic development. I think this is a unique perspective. As an engine for economic development, principals need to develop deep ties with employers and understand the business climate of their community. We did at Wheeling High School. We found that our community had the third largest concentration of manufacturers in the entire country.

There was nothing the school was doing to help meet the needs of that particular employer group or that industry. In my meetings with employers, they stated, "We have a shortage of qualified individuals applying for and even considering this as a well-paying career." There seemed to be a mismatch between these local opportunities and the fact that many of our residents were living in poverty. The school could play a role in making the connection to local employment more directly.

The initial efforts were to respond to the immediate need. Keep in mind that at that time we were still in the midst of the Great Recession. A lot of people thought I was crazy for trying to create a manufacturing lab at a time when everyone was saying, "We're supposed to be only about college-ready or college-bound students. Why are you going backward instead of forward?" Where I did get support was from the Economic Development Office in the Village of Wheeling. They have all of these manufacturers who they're trying to serve and support, and they understood if they could not deliver on a talent pipeline locally, they could consider relocating. Their encouragement and validation helped me move forward.

EDUCATIONAL FOCUS – TEST SCORES OVER CAREER READINESS?

Our entire public education system, from a lens of accountability, focused on trying to ensure that our schools are graduating students who are ready for college. This was a result of a loss in our belief that the earning of a high school diploma actually meant something. A multitude of assessments were introduced, and the focus of the entire K-12 system narrowed for over a decade on testing and test scores. And while that is one valid measure, it is not the only measure of student potential for success. In that process, we got rid of everything and anything not directly contributing to test scores, and the central focus of school leaders has been on, "Can my kids score a certain number on a standardized test?" So you're correct in the sense that the end goal is an assumption that if we can get all the students high enough test scores, all of them are going to college, and that is the right path. However, as any parent knows, students are more than just a single test score given on a single day. "Redefining Ready," adopted by the National Superintendents Association, attempts to counter this assumption by redefining the way we should be holding schools accountable. Utilizing multiple measures, students can show readiness for both college and career.

CONSTRUCTION – A NEW EDUCATION MODEL

We've structured the comprehensive high school academic program, and all coursework, around each of the sixteen national career clusters along representative career pathways. Architecture / Construction is certainly one of them, where students have designed and built a house from an empty lot. This year we transitioned to flipping houses. There are lot of opportunities in home remodeling and many of our students are interested in running businesses in that field.

As we transitioned to flipping houses, we founded 214 Studios, where we have students filming the house flip, similar to what you'd see on TV. We've called that series, "High School Flip." We are trying to give students who are interested in media, a real-world production experience in writing, filming, and editing. Whether it's the students who are tearing down the walls or behind the camera filming the work, each experience affirms their career choice. Is this what they want to do with their life? We should make sure that high school is the place where they're able to discover that future.

NASA! HIGH SCHOOL STUDENTS SHOOTING FOR THE STARS!

The application of manufacturing challenges students in math and science beyond the theoretical to real-world projects. Wheeling High School was selected as one of fifteen schools in the nation that prepared components for the International Space Station in partnership with NASA. The tolerance for manufacturing the parts were less than a human hair. There is no room for error when you're working with NASA. The application of skills, technical and interpersonal, to actually produce these brackets and handles with real world consequences was incredible. That project, that will impact the students and teachers involved for a lifetime and provided learning to them, is applicable to other teams and industries.

ROBOT RUMBLE

The process of developing a product is also replicated through our "Robot Rumble." Students begin with a hunk of metal. They design and build remote controlled robots that are going to combat other student-built robots at a regional competition. The result is that bringing all these experiences to scale gives students a real opportunity to authentically discover their future.

AFTER YEARS OF GROWTH, MATH PROFICIENCY OF U.S. STUDENTS IN 8ᵀᴴ GRADE DIPS

Percentage at each achievement level of the National Assessment of Educational Progress (NAEP)

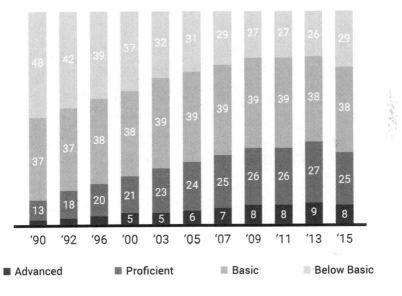

'90 '92 '96 '00 '03 '05 '07 '09 '11 '13 '15

■ Advanced   ■ Proficient   ▨ Basic   ▨ Below Basic

Source: NAEP Data Explorer, National Center for Education Statistics

HIGH SCHOOLS NATIONALLY LEADING THE WAY FOR OUR STUDENTS

Generally speaking, education in this country has failed many of our youth. We cannot change the past and what has happened. Our math and science scores as a country are amazingly low for the educational culture that exists in the U.S. While I think that other countries certainly measure their population differently than we do, there still is a problem that we need to address. China evidently only had four provinces participate in the PISA 2015 study, and as a result it is not included in the results.

One of the biggest cross-national tests is the Program for International Student Assessment (PISA), which every three years measures reading ability, math and science literacy and other key skills among fifteen-year-olds in dozens of developed and developing countries. The most recent PISA results, from 2015, placed the U.S. an unimpressive 39th out of 72 countries in math and 24th in science.[4]

We have been teaching to a test and allowing different states to define their own (and different) standards. Students are passing varying tests and not developing the critical thinking and problem solving skill sets. These are imperative skills for manufacturing related jobs. A percentage of students who are entering into engineering coursework are saying that they can't make the grade and transferring into different majors. This could be a direct result of the high school preparation in both math and science. If the level of performance was appropriate, engineering students would find themselves prepared and graduate into engineering careers. High schools also find themselves being judged by only what percentage of students go to a four year colleges and take and pass advanced placement courses. In spite of all that, the Society of Manufacturing Engineering is making a difference with Project Lead The Way. High schools that are participating are making the connection in project based learning. How can students learn math effectively unless they understand how they can use it in real life? What about being able to relate physics to applications that bear a cause and effect mentality?

CARDINAL MFG IN WISCONSIN – IMITATION IS THE SINCEREST FORM OF FLATTERY

One of the truly great educational stories in the country is a Wisconsin high school where they have formed a manufacturing company. This business model teaches life lessons which go well beyond the school years that its participants experience. It changes their lives for decades to come.

Cardinal Manufacturing began in the Eleva-Strum School District during the 2007-2008 academic year when instructor, Craig Cegielski, approached the school board about the potential of pursuing an in-school manufacturing business similar to one he started in his prior position in the school district of Antigo, WI.

The school board approved, and since that time, Cardinal Manufacturing has gone from its infant stages to a company with significant annual sales and national notoriety. The growth of the program has attracted national and international attention, and Cardinal Manufacturing has attended national tradeshows and hosted celebrity guests including Wisconsin's governor, Scott Walker. Cardinal Manufacturing has served hundreds of customers, from private individuals to clients throughout the state of Wisconsin and other parts of the country. A number of students have gone directly to skilled employment positions after high school, but most choose to go on to post-secondary education through technical college or the university system. Chippewa Valley Technical College and UW-Stout have taken a particular interest in the program. Cardinal Manufacturing has also built strong relationships with a number of private companies and professional organizations which have been supportive through donations, advice, publicity opportunities, and projects.

In-school programs such as Cardinal Manufacturing serve as a grassroots economic development effort. Not only do these programs expose students to career opportunities in manufacturing and teach students soft skills for future employment, but they also work toward changing the attitudes of counselors and parents to be more open to the idea of encouraging students to look at manufacturing careers. Students get hands on opportunities to try out these roles before making an expensive decision in choosing a post-secondary program. In other words, kids get the chance to try welding, machining, construction, production management, accounting, office management, and marketing prior to committing to a major or area of study. The services provided through the program are worthwhile and valuable to the customers who pay for the service. Since the very beginning, Cardinal Manufacturing has been self-sufficient, meaning that they have not ever requested a special budget from the school district.

The instructor and creator, Craig Cegielski, talks about the dozens of schools that come to learn more about his business model for the school program. Two schools in the midwest have started their own versions. Rocket Manufacturing (Rock Valley, IA) and Cub Manufacturing (Madison, IN) both have paved their own paths after being shown the way by Craig's Cardinal Manufacturing.

Each year, many teachers and administrators contact Cardinal Manufacturing with the desire to learn about the program and how to create something like it in their own school districts. Cardinal Manufacturing is a school-based enterprise, a real student-run manufacturing business. This one-day workshop will provide you with a guidebook, hands-on opportunities, and firsthand experience with an existing and successful program. You will learn how Cardinal Manufacturing began and the steps that have taken place since then to grow it to where it is today. Meet student employees, learn about administrative concerns and management, see our facilities, meet some of our business partners, and feel inspired to create your program. Not only will you learn about our program, but you will have the opportunity to work one on one with our administration, service providers, and business partners to help begin drafting a work plan you can put into motion right away.

My hope is that this model finds its way to schools all over the country, in each and every state.

FINDING AMERICA'S GREATEST CHAMPION NOW!®

One goal that I have is to have a CHAMPION Now!® poster and magnet in every Project Lead The Way or PRIME Program in the country to raise awareness for manufacturing careers and also opportunities in this country. I would like to think many copies of my book would be in schools all over the country to give not only students, but teachers, administrators, guidance counselors and others, access to concepts that can change young people's paths to becoming their own Champion! Just having these posters in the manufacturing labs will be a perpetual resource as new students come in each year.

My dream is to help open a high school, or even a lower school that promotes and encourages knowledge in the manufacturing and engineering fields. This school would feed the curiosity young people have for something that they have a love for. There would be an entrance exam to ensure that people entering into this field would have the adequate math and science aptitude to succeed. In this country, our math and science scores are not indicative of our place with the rest of the industrial nation world. What happened? How can we allow such mediocrity? This from one of the world's leading nations in manufacturing, productivity, innovation and creation.

Among the thirty-five members of the Organization for Economic Cooperation and Development, which sponsors the PISA initiative, the U.S. ranked 30th in math and 19th in science.[5]

Countries far less advanced somehow repeatedly outscored our students. This is something, going forward, that should change. Admittedly, many say that the measurements are not an apple to apple comparison. We still need a national education wake up call. Our manufacturing leaders have to stop accepting mediocrity and expect excellence. It is this that will allow us to once again rise in manufacturing and engineering superiority.

HOW THE U.S. COMPARES ON SCIENCE, MATH AND READING SCORES

Score rating of 15 year-olds taking the 2015 Program for International Student Assessment

	Science	Math	Reading	AVERAGE
Singapore	1	1	1	**1st**
Japan	2	5	8	**5th**
Canada	7	9	3	**7th**
South Korea	10	6	7	**8th**
Taiwan	4	4	23	**10th**
Germany	14	15	11	**13th**
Norway	23	18	9	**16th**
Switzerland	17	7	28	**17th**
UK	17	26	21	**21st**
Sweden	27	23	19	**23rd**
USA	24	39	24	**29th**

Source: OECD, PISA 2015. PEW Research Center.
Results from China not included because only four provinces participated in PISA 2015

THE COMMUNITY COLLEGE CAREER TRACK
BY THOMAS J. SNYDER

In reading about Thomas Snyder and his career at Ivy Tech, an Indiana institute known for technical education excellence I learned of his commentary and passion about the community college's alternative path to success for its students.

Everything is different now. I don't mean in relation to college costs but to the employability and lifetime earnings potential. In fact, the sticker price for a bachelor's *degree in a Indiana public residential four year college is $72,000.*[6]

I have found over the decades working in the manufacturing sector that Snyder's comment about employers being willing to pay for night class training for their employees to be spot on, and true facts. It has been my experience that this is a win win for both parties.

Nothing can hide the fact that community college is the smart higher education choice for an increasing number of students. Professional certificates and associate degrees have become the favorite gateways to many of today's and tomorrow's best jobs.[7]

Some of Snyder's commentary adds that going to community college in lieu of the first two years of a four year university is a viable alternate path. He goes on to say that a degree from an elite college may not hold the same mystique and impact it once had.

Conversely, there is strong evidence that community college can accelerate your bachelor degree attainment and/or quickly put you on a desirable career path. Not for the first time in the nation's history, the American dream stands in need of reinvention and renewal. The process of reinventing and renewing the American dream will be a complicated one with many elements, but there is no doubt the community colleges will remain at the center of the story over the years ahead.[8]

SMART KIDS CAN GO INTO THE TRADES TOO

In March 2018, in an article from the Wall Street Journal, Doug Belkin writes about parents having the dilemma of rising student debt and businesses pushing faster and cheaper paths to the workplace. Charleroi (PA) High School student Ralee Nicholson exudes college ready qualifications with honors class A's and upper percentile college board scores. Despite this, and guidance counselors', teachers' and other adult objections, Nicholson hopes to study in a technical college diesel technician program rather than a conventional four year college path. Many influencers prefer this more expensive educational model, as this has always been the preferred and accepted choice. The other path was always considered for kids who were either not able to make the grade or those in trouble. This is a partial reprint of Belkin's article.[9]

The conversation is being fueled by questions about the declining value of a college degree as well as the rising cost of tuition and student debt. Low unemployment and a strong job market are exacerbating an already growing skills gap, raising prospects for tradespeople, like welders, who are in high demand.

In 2009, the last year for which data is available, 19% of high school students were concentrating in vocational subjects, down from 24% in 1990. Even as more students enroll in college,"40% to 50% of kids never get a college certificate or degree," said Tony Carnevale, director of the Georgetown University Center on Education and the Workforce. And among those who do graduate, about one-third end up in jobs that don't require a four-year degree. This has prompted a rethink about the value of colleges and is fueling a separation between the winners and losers in higher education.

These forces are leading to a course correction now rippling through U.S. high schools, which are beginning to re-emphasize vocational education, rebranded as career and technical education. Last year, forty-nine states enacted 241 policies to support it, according to the Association for Career and Technical Education, an advocacy group.

"We've got to make sure we're sending the right signals and also preparing people for the world as it really is not as maybe we'd like it to be," said Pennsylvania Governor Tom Wolf.

Dr. Carnevale calls the movement a "counterrevolution." But he also believes it will remain a hard sell, particularly in affluent suburbs and for high-achieving students.

"Raelee was smart from the time she was a baby, from the time she was two nobody could dress her, she was always a leader and she had her own mind," said Raelee's mother, Beth Nicholson, a nurse. "I always expected her to go to a four-year college. That was my expectation."

"I really like working with my hands." She doesn't listen to those trying to dissuade her from her passion. "Diesel mechanics charge $80 an hour," she says.

DO THE MATH...

This is a sample of two students. One goes to college for 4-5 years taking loans or using money other than a gift from Mom and Dad. If we use Raelee's financial model with diesel mechanics $80 per hour wage, that would equate to over $150,000 a year earnings.

4-5 year college degree

	Tuition/Income	cost/payback
Year 1	$(50,000.00)	
Year 2	$(50,000.00)	
Year 3	$(50,000.00)	
Year 4	$(50,000.00)	
Year 5	$(50,000.00)	$(250,000.00)
Year 6	$75,000.00	$15,000.00
Year 7	$75,000.00	$15,000.00
Year 8	$75,000.00	$15,000.00
Year 9	$75,000.00	$15,000.00
Year 10	$100,000.00	$20,000.00
Year 11	$100,000.00	$20,000.00
Year 12	$100,000.00	$20,000.00

$(130,000.00) NET

2 year vocational degree

	Tuition/Income	cost/payback
Year 1	$(20,000.00)	
Year 2	$(20,000.00)	$(40,000.00)
Year 3	$50,000.00	$10,000.00
Year 4	$50,000.00	$10,000.00
Year 5	$50,000.00	$10,000.00
Year 6	$50,000.00	$10,000.00
Year 7	$75,000.00	$15,000.00
Year 8	$75,000.00	$15,000.00
Year 9	$75,000.00	$15,000.00
Year 10	$75,000.00	$15,000.00
Year 11	$75,000.00	$15,000.00
Year 12	$75,000.00	$15,000.00

$+90,000.00 NET

- *What this shows is that a 4-5 year degreed person in year 12 (seven years after graduation) is still $130,000 in debt.*
- *The 2 year vocational student in year 12 is in a surplus of $90,000.*
- *This is almost a quarter million dollar difference between the two.*
- *This assumes that only 20% of a person's wage can be used to pay back principal, while the vocational graduate makes 25% less per year, which may not be the case at all.*

If you include the financial model to take the $250,000 gift, deducting the $40,000 and instead invest the balance of $210,000 in a conservative investment, then the justification takes an entirely different level of justification. Now the 2 year student has enough money to start his or her own business. At a rate of return of 5% re-invested, the 2 year student would have $325,800.00 to fund a startup, pay cash for a house, or re-invest in stocks. Add the $90,000 to that, and you are approaching a half a million $ nest egg for someone who is quite young and only went to a trade school for two years.

HIGHER EDUCATION: HOW THE U.S. COMPARES GLOBALLY

In this book I make a number of comparisons to other industrialized nations. I compare the culture specifically to both Switzerland and Germany and others. When making comparisons about our cultures, it doesn't take long to wonder how we also compare academically. In looking at PEW Research and college graduation rate we find interesting numbers.[10] While we score higher than Germany and Switzerland in these numbers, this doesn't account for account for those populations that enter into a robust apprenticeship program in both countries.

According to the Organization for Economic Cooperation and Development's "Education at a Glance" report. U.S. college graduation rates rank nineteenth out of twenty-eight countries studied by the OECD. What can we learn from this finding? First, we have to understand the term "educational mobility." In an article published in Reuters in 2014, higher education levels are associated, not just with higher earnings, but also with better health, more community engagement and more trust in governments, institutions and other people. The OECD report found that in 2012, 39% of young Americans were expected to graduate from college, compared with:

- *60% in Iceland,*
- *57% in New Zealand*
- *53% in Poland*
- *39% in the United States*
- *35% in Canada*
- *31% in Germany*

- *31% in Switzerland*
- *29% in Spain*
- *27% in Turkey*
- *26% in Italy*
- *23% in Chile*
- *23% in Hungary*
- *22% in Mexico*

The article goes on to state that "about half of young people in OECD countries have at least matched their parents' level of education. But in the United States, a larger-than-average proportion had less education (so-called downward mobility) while a smaller-than-average population had more education (upward mobility)." In 1995, the United States was the frontrunner on OECD's list with a 33% graduation rate.

A CONVERSATION WITH VINCE BERTRAM
PROJECT LEAD THE WAY – "DREAM DIFFERENTLY"

When I got involved in technical education, I learned that high school technical education was being lead by a project based learning organization. This is Project Lead The Way (PLTW). Over the years, I have met with many high schools on the PLTW program. I interviewed the president of this inspirational organization—Vince Bertram. He educated me on the program's development and reach:

Six years ago we had 300,000 students participating in PLTW programs in about 2,500 schools. This year we're almost in 11,000 schools with millions of students participating. What we try to do is create a scalable and sustainable organization so that we can deliver this kind of learning experience to millions of students across the country.

Our vision is to ensure that every child in America has access to PLTW programs. Not that every student will take it, but that they have access, that we eliminate barriers to access regardless of where a student lives or their economic situation.

We try to provide real world application and relevance for students. We want our students to understand why math is important, why science is important, not just to learn math or to learn science. These are tools to solve problems. All of our courses are applied, problem based type of courses. As early as in pre-kindergarten, we can give them an entirely different type of experience.

We also provide extensive professional development for teachers across America. This past year we trained over 12,000 teachers on how to teach in an activity problem based classroom. When teachers learn to teach in that manner, we can take school and class and move it from a place of boredom and isolation to one of collaboration and engagement and excitement for students.

We want students to have great careers. We want them to develop the skills and knowledge that will allow them to pursue the careers that will give them an opportunity for economic prosperity.

Right now, no one is really accountable for student outcomes. It's graduation rates. Even in higher education, it's graduation rates, and college indebtedness. We don't ask how are our students are doing in careers. How many are employed within the majors that they are pursuing? What are the financial outcomes—not just at the time—within six months, or six years or ten years out?

Is the investment in education delivering those kind of outcomes? I would really like for us to get to those answers and align accountability, not to what I consider both premature and insufficient measures of performance, but to more enduring type of outcomes.

When you think about the conversation around culture, and really the future of our country, we have to take responsibility for our students today. We have to help them understand the opportunities available to them and what it takes to pursue those opportunities. I think your direction, the focus on those kind of outcomes, is very important in how we prepare our students for the workforce. If you've seen my latest book, it's called Dream Differently: Candid Advice for

America's Students. I'm making the argument, as well, that this is about skills, and it's about alignment of your interests with those skills that allow you to have enduring careers.

A CONVERSATION WITH WAYNE LARRIVEE
"LET'S NOT PIGEONHOLE OUR YOUTH"

In Chapter Six, I talked about Wayne and I coaching together. Years ago, Wayne and I decided to go with our sons to an overnight baseball camp in Indiana on the finer arts of hitting, among other things. It was held in a former monastery facility, and we all had a great time together. As we got to know one another, we became good friends, and to this day, I look forward to hearing from Wayne and his wife Julie. During our time together coaching baseball, it became evident to me that the Larrivee's shared some key values as parents that I felt were awesome in raising children in today's culture. Here Wayne speaks to the decision pertaining our children's post high school education:

I think that we got into a situation—with maybe our parents and then us—kind of always focusing on the four-year education rather than the trade jobs. A lot of kids shouldn't be in college. They really have no business being there. They have no real interest in being there. They don't get a whole lot out of it.

They come out of college, and they don't know what to do. They've got a degree that will do them no good at all. We're just now starting to see where it's very important that everybody consider the trades. They've been looked down upon for a generation and a half. This is where manufacturing comes into play. This is what this country was built on. There's nothing wrong with going to a trade school, getting a job that you can hold and be proud of, work at, and make good money at for your career. In fact, many kids, not all, probably are better off aspiring to that.

I'm seeing that's starting to happen in this country, maybe more out of necessity. College has just become so expensive. I just look at what these kids are coming out with, $100,000 or more dollars worth of debt. We're all sitting here wondering why the housing market is so soft. The fact is that the next generation can't afford to get into a house until they get this debt they have paid down.

I mean I don't know how these kids do it. They have $100,000 worth of debt, and they're going for a job that's maybe paying $35,000 to start. I don't know how that works. That's why kids default on loans. That impacts their credit rating. It's a mess. It spirals as you go through society and you go through the years.

It starts really with the approach that parents take. Everybody steers their kids toward a four-year school. Make sure you do well in grade school and high school. Pass that ACT test and get a high mark there so you can get into a four-year school.

Parents have got to understand their kids. I got a kid here who doesn't quite know what he (she) wants to do. Maybe he's (she's) really good with mechanics. Maybe he's (she's) really good with working on machines. Parents have to make sure that these kids know that they don't have to go to a four-year college to be successful. In fact, they've got a better chance of succeeding early in life if they don't go to a four-college in many respects because of the economic impact I talked about a moment ago. Part of the reason why the middle class is disappearing is because of exactly what you're talking about here with manufacturing. You need a strong manufacturing base. That's the middle class. That's it in a nutshell.

Maybe you should go into the military. Maybe you should do something until you figure out what it is you want to do—then go for it. Chances are that it's not going to require a four-year college degree. I think that would probably help your business and manufacturing in general. If kids came out of high school and they understood that they can go to this technical college and get their career going.

I think that kids should be allowed to not be pigeonholed. We pigeonhole them in sports all the time. It's done very poorly on the youth level. They're targeted to be one thing, and they're this then. When you and I grew up, kids played more than one sport. They played three or four sports. Quite frankly, today they're kind of pigeonholed in sixth grade. They're targeted in travel teams, and they're only doing one thing. That's kind of a sports analogy to what is happening and is more important in life. I think that a lot of times we, as mentors or parents, may determine (or try to determine) what this kid is going to be and wants to be. What we need to do is foster an environment where that child has a chance to experience for himself (herself), decide what he (or she) is, where he (or she) is going, and what will make him (her) happy in a career and in life in general.

Wayne's commentary leads to an entirely new concept of concern for the parents of this country. Are we doing our best? Are we giving the next generation the best we have?

A CONVERSATION WITH GARY SKOOG AND THE GOLDEN CORRIDOR

I have been involved with Gary and the Golden Corridor for a number of years. As a group, we approached President Ken Ender at Harper College on the importance of revitalizing the manufacturing program with the Chicago Metropolitan Agency for Planning (CMAP) Marketing Project. Harper made a significant focus on manufacturing technology with the support of the Fabricators & Manufacturers Association (FMA). Gary and I continue to keep in touch and be connected in the promotion of manufacturing in every way we can. Gary is truly a manufacturing Champion.

Our educational system is really behind. There are other global educational systems where, early on in high school, you start working. That doesn't mean you have to declare your lifelong career, but you do make a choice, early on, and you start rotating being at work and school.

You can earn and learn. This is a great concept. I think we're much too focused on the notion that everyone has to go to college. Statistically, it just doesn't add up that we have 1.3 trillion dollars in student debt. In six years, maybe 50% of the people who enrolled in a four year institute graduate! Not too many manufacturers would make it if their failure was 50%. A lot of those kids come out, and they've got a big debt. They don't have a degree, and they have trouble finding a job. I think we need to start much earlier with students with career exploration. This is tough as many students are just not focused on a career.

If our educational system can change where people get hands-on, work-based learning during their sophomore, junior and senior years, they're completely keyed into, "I am going to go on to a four year degree," or "I think I can go right into an apprenticeship that will give me an AA degree. A lot of them are hooked up with community colleges. Later on, if I want to go into being an engineer or a manager, I can finish out my school, and there's a good chance my employer will pay for it." You see a lot more of these apprenticeships popping up and some of them are just totally driven by the company itself. Some of them are Bureau of Labor sanctioned so that they can get various certifications. I think they are a great thing.

Gary's commentary fittingly leads to the question on the return on the investment on a four year degree. Is there a need for a change in culture? What is the job market wanting? Does everyone need—and can they afford—the four year university experience? Too many people go off to college to figure out what they want to do with their lives. We, as parents, are part of the problem. We think that we are giving them "space" to make the right decision.

Instead, what might be much more effective is to have our young people put thought into what they feel they can be successful in, then make the decision or path for that career. The cost of college and the entrance requirements have become such that more careful thought should be put into each of our children's passion in life. We need to look at the full ROE. In the next chapter, we will look at Florida and Wisconsin, where people and organizations have a focus on manufacturing careers.

LEADING THE WAY: FLORIDA AND WISCONSIN

"... we need to make sure we have a conversation and force the issues with those in education to take a look at people who are good with skill sets and with their hands. I am a big proponent of education, but it needs to be dual in nature. We need to make sure that we are giving all of our students good opportunities."

– Lake Ray, President of First Coast Manufacturers Association

Both Florida and Wisconsin, in different ways, have empowered manufacturing companies and education. Wisconsin is prototypical manufacturing employment-based, and Florida is not but has advocated manufacturing careers and will draw manufacturing companies to their state as a result. Wisconsin has a robust manufacturing community and state economy. Sixteen percent of their workforce is employed in manufacturing positions.[1] Despite all of this, manufacturing companies complain that they cannot find enough good skilled employees.

It was longtime friend, John Stilp (from Milwaukee Area Technical College), who invited me to the FLATE organization in Florida. He invited me to participate in the National Visiting Committee for this organization. We meet once a year, and we report on the findings of their efforts over the last year in carrying out their mission statement. They have technical colleges throughout the state using a two year degree that is named an Engineering Technology degree. They use MSSC (Manufacturing Skill Standards Council) as a stackable credential program that allows students to move towards a degree that manufacturers can use to differentiate applicants in their quest to find more notable hires. This is a great model to use across the country. Marilyn Barger and Richard Gilbert are the leaders of this program but are quick to point out that they aren't the feet on the street to carry it through.

FLATE has developed a FLATE and Made in Florida website. They have designed coloring books for elementary students with images in manufacturing settings and explanations on their relevance. They conduct many factory tours with high school youth. All of this is a tremendous and effective outreach for the manufacturing sector.

Admittedly, Florida's manufacturing economy does not rival that of many other industry driven states. In spite of that, they have perfected a task force that gets the message out around the state. In their participation in Manufacturing Day®, they have consistently led the country as one of the top five states with the most events for that event.

For my entire career, I have witnessed what a community technical college should look like in the states of Wisconsin, Illinois, Indiana, Florida, Pennsylvania, and others. What I have noticed is Wisconsin's commitment to the manufacturing industry. For decades, the state has implemented some of the most advanced CNC manufacturing labs. I don't comprehend how so many other states can fail in comparison to the standard set by Wisconsin. Lawmakers, politicians, and representatives have listened to business leaders and manufacturers and supplied them with labs that not only have state of the art machine tools, but talented instructors as well. How can they accomplish things that other states can only dream about?

We should all challenge leaders in other states to step up to the plate and form partnerships with visionary manufacturers in their areas. We can no longer afford to have the technical college that does not offer the advanced training for our manufacturing companies to be on the cutting edge of technology. At the same time, we cannot fill these facilities with the best in machine tools and NOT have students to fill the classes. It takes a collective approach.

Florida has about 4.2% of its working population in manufacturing.[2] Here is a state that is known for its tourism industry, doing its best to attract the next generation of workers who will bring more industry to the state. In recent years, extreme weather has challenged the state. Tourism has taken a deep hit in reduced travel to the sunshine state. Meanwhile, Florida has so many workers from the north wanting desperately to relocate to a warmer climate if they could only find a good paying manufacturing job like they had in other parts of the country. Economic development groups from all over the state look to train the next generation's manufacturing workforce. This is what will bring the almost 21 million people in the state of Florida,[3] to the next level of economic independence: that which is not dependent on the sun filled skies alone.

A CONVERSATION WITH MARILYN BARGER
EXECUTIVE DIRECTOR OF FLATE

FLATE is funded by the National Science Foundation Advanced Technological Education (NSF ATE) program for community college technical programs. It focuses on two things: more and better prepared technicians for the advanced technology industries in our country. We wanted to do this for manufacturing in Florida. We wrote a proposal for our center which now has been funded since 2004. We worked closely with industry, colleges and the Florida Department of Education to develop and implement manufacturing education pathways that start in middle school and goes through college. Today, the A.S. Engineering Technology program is offered in twenty-three of Florida's twenty-eight community and state college with over 2,100 students enrolled. Most graduates go directly to work as manufacturing technicians. The college and the growing number of high school programs across Florida work together to support our manufacturers' long and short term workforce needs.

In 2013, we realized that the new national Manufacturing Day® was an opportunity to support the two year programs with outreach to the K12 students. We have had a history of taking students on industry tours and built our Manufacturing Day® - Florida efforts on that success. Our goal is to expose as many secondary students as possible to advanced manufacturing during highly visible tours. Coordinating this effort statewide allows all these local and regional activities to be amplified.

We reached out to school districts and manufacturing organizations, asking them to participate in student tours for Manufacturing Day® - Florida. With help from district offices, we matched schools with companies and provided logistic support. It actually became the "thing to do." We provided regions a "how to" guide and a number of resources and suggestions for funding. The program has continued to grow, with nearly 5,000 students annually participating statewide.

The outcomes that FLATE has been most interested in is seeing what impact the tours were having on students. We developed a survey for the students who go on Manufacturing Day® - Florida tours. We administered those through the tour host, the schools, and/or the regional associations. We tally the survey data to learn how students are impacted by the experience. We have been surveying students since 2013 and reporting back to participating regions. Additionally, these tours have fostered many new partnerships between businesses and education. We hope those partnerships will continue to grow and that students exposed to manufacturing on tours and in their classrooms will consider manufacturing as a good career option and pursue it via our education pathways.

We hear from students, teachers, manufacturers and other stakeholders often that the tours are "awesome." I think it's really important for students to get exposure to the workplace. It is hard for them to imagine what people do every day at "work." We hope they will see something that they like in the exciting and fast-paced manufacturing world. It's also important to have a good strong educational pathway with lots of options for students. We have a great career pathway system in Florida and FLATE and its partners are always working to make it better for manufacturing careers.

A CONVERSATION WITH LAKE RAY
PRESIDENT OF FIRST COAST MANUFACTURING ASSOCIATION

In 2012, I was asked by the West Nassau County Economic Development Council to speak at an event. I was there speaking, with the CHAMPION Now!® message, about manufacturing and the tremendous opportunities that manufacturing careers hold. During the course of that day, I met two fascinating people after the gathering. One was Lake Ray who was a Florida State Representative. Lake is now the president of the First Coast Manufacturers Association and works tirelessly to hold the manufacturing banner high for northern Florida.

The state of Florida is beginning to engage in the manufacturing sector. Roll the clock back about ten years ago, and you would see that the state's economy was focused on tourism and land development. Strong economies are built on three things:

You make it. You mine it. You grow it.

The state's economy was broken because we were not making things and mining power was increasingly limited. Agriculture, however, was strong.

I entered the legislature in 2008, when the economy was upside down. How could you broaden the economic drivers for Florida? I focused on ports and infrastructure for moving goods. Florida has fourteen deep water ports. We needed a better system for moving manufactured products. Moving products, for assembly or goods for consumption, requires a good transportation system.

The first focus was on how to improve the ports and the second step was identifying the stakeholders. The stakeholders were primarily manufacturers and distribution companies like Amazon and others.

The next investment was manufacturing. We wanted to import and export products. We developed legislation that would enhance our manufacturers, making investments and policies to lure companies to Florida. A key policy was elimination of a sales tax on manufacturing equipment. For a company investing $10,000,000 in a project, saving sales tax would be a significant reduction in costs, making Florida more attractive.

We have our universities focused on higher levels of education and technical training. We have a state college system that is more focused on the associate degrees in manufacturing, associate in science, or associate in arts degrees. They work with companies to provide and develop more focused program targeting skill set the companies require for their employees.

The college state system is more dedicated towards the ultimate online worker who actually manufacturers things. We are beginning to see breakthroughs in the public school system. The legislature began mandating STEM (science, technology, engineering and mathematics) curriculum. We're beginning to see some signs of students coming out with their diplomas who have minimal skill sets that are attractive for employment. One of our companies, Johnson and Johnson, is working with our school system to identify the future workforce. Targeted students can attend high school, alter their school work and intern at Johnson and Johnson. The program teaches them the desired skills, letting them operate equipment and work with the staff.

The students are trained in the process. The company decides who will continue the program. The coursework includes our state college at Jacksonville. Upon high school graduation, a special associate's degree in manufacturing can be earned in one year. They are typically employed at an active salary of about $65,000 a year by Johnson and Johnson with the degree.

Our new state policies have reflected about an 80,000 job increase in the manufacturing sector. In 2010, we were approaching a low employment level of around 300,000 manufacturers. Today our employment of manufacturers is approaching 400,000 workers representing about 5% of the state's workforce.

Aerospace is growing on our space coast and in Central Florida. We also have a lot of other companies like Northrop Grumman who manufactures aircraft components.

We have a number of manufacturers that focus on entertainment. Others support making products for Disney and theme park robotics. In Jacksonville, we have a number of companies in food processing such as White Wave who make almond milk. We have glass makers and aluminum can makers that make metal bottles.

We have other goods that are familiar to many such as Anheuser Busch and Bacardi bottling company. We have Johnson and Johnson, formerly Vistakon for eye contacts and Medtronics, who make surgical medical equipment including pacemakers.

Education and the workforce is becoming a greater consideration for keeping and attracting manufacturers locally above incentives. Companies looking to relocate ask, "Does the population have the skilled workforce we need?" If not, "Is the training available?"

Education must refocus on people who are good with skills, not merely college degrees. I am a big proponent of education, but it needs to include vocational and high tech training. We need to make sure that we are giving all of our students good opportunities.

PUBLIC HIGH SCHOOL PRE-ENGINEERING SUMMER CAMP

Even though I now live in Chicago, I travel back to Jacksonville often. While watching a local Jacksonville TV program, I learned of a group called Renaissance JAX. Their vision is to introduce LEGO robotics to every student in Duval County. This organization was founded by a young man by the name of Mark McCombs. He is a very energetic and inspiration individual who wants to bring manufacturing technologies to young people under the age of fourteen.

"I want to spread manufacturing skills and mindset to K-12 students in all of Northeast Florida. (We manage the fastest growing high school robotics league in the state attached to $80 million in scholarships.)" - **Mark McCombs.**

That same day I met one of the GE engineers who recently moved a facility into Jacksonville, he indicated that they were donating a CNC vertical machining center to their cause. Mark was over the moon and excited that Renaissance Jax would have their very own CNC machine! One of the exchanges that Mark and I had was that I did not want to compete with his organization if I brought something to Jacksonville.

"Renaissance Jax builds interest for future manufacturers and CNC programmers by building the FIRST Robotics program at scale so kids have some fundamentals that make learning and pursuing manufacturing much more viable compared to what is available in the region. Because of the rapid growth of students with robotics skills in Northeast Florida due to Renaissance Jax, there is an even

*greater need for CHAMPION Now!® to tool up their program to take students to the next level." - **Mark McCombs**

CHAMPION NOW!® CNC ROCKS™ SUMMER CAMP

Through the help of the leaders at Frank H. Peterson Academy in Jacksonville, we were able to develop a one week "CHAMPION Now!® CNC Rocks™" summer camp program. Through many conversations with Frank Peterson, Principal Jessica Mastromatto, Assistant Principal Christine Bicksler, and CTE Instructor Russ Henderlite, I determined that this was a worthwhile project. Iverson personnel programmed the machine, set up the jobs, trained the instructor and packaged the machine for shipment. One of my CNC machines and a computerized inspection system was shipped down to their facility.

Russ and I introduced nine students on how to manufacture components and check parts made on the machine. These young people were shown a whole new world that they knew of, but had no tangible point of reference. I cannot describe the inspiration I feel when a young person listens to your every word in hopes of using the information to make a life-changing decision that will set him or her on the path to success.

Fascinating enough, during the camp, I met the new CTE director for Duval County Schools, Ryan Rewey. Ryan came from Wisconsin where the support for technical education is significantly greater than most other states in the country. We knew a lot of the same people in the industry, and both of us are motivated to bring workforce development to a new level in Jacksonville, Florida. The Frank Peterson leaders had this to say about the experience:

"From the CHAMPION Now!® CNC Rocks™ camp I have had students who have been motivated to volunteer at local CNC shops. The program allowed students to experience the design, manufacturing, and verification processes first hand."
*- **Russ Henderlite Instructor**, Advanced Manufacturing Technology, Robotics*

"Bringing the CHAMPION Now!® CNC Rocks™ camp and experience to FHP (Frank H. Peterson) benefits students by allowing them to see the opportunities available to them in manufacturing and future career options."
- **Ms. Mastromatto** *Principal at Frank H. Peterson Academies with Duval County Public Schools*

"The CHAMPION Now!® CNC Rocks™ training provides relevance to learning. Mr. Iverson has a passion for helping students achieve success."
- **Ryan Rewey** *Director of Career and Technical Education Duval County Public Schools*

"Developing industry partnerships like with CHAMPION Now!® and Iverson & Company is crucial to prepare students for careers beyond high school."
Christine Bicksler Assistant Principal, High - **Frank Peterson Academies**

Some of the campers gave me commentary that was very enlightening. These young people were so interested and engaged in the event, and their commentary proved that they each truly got significant value and knowledge from this four day camp. I cannot tell you how gratifying helping these young people is and giving their future new possibilities.

"I did not know what a CNC machine was. Now I fully understand what it is and what it does and how you get a job. I didn't know that you could get a job from this. I am already mind blown with just the basics!" - **Cadeem Adams**

"When you go to college, make sure you are going to do something you really want to do, or something you can take into the real world with you and still make a living off of it and you also enjoy it." - **Jamir Brown**

People like Mark McCombs need their stories told. More people need to be inspired to mentor, inform and educate those who may not know what engineering or manufacturing is in the world. Once they know about it, they can decide if it is or is not their mission and role in their life and or career. Unfortunately right now, our culture does not have enough exposure, emphasis or understanding of either in order for our youth to have enough data to have this on their radar.

One of the biggest flaws in engineering today is that too many engineers come out knowing how to design but not a clue on how to make the products that they design. This results in one of two scenarios: either the product that they have engineered can be made, but is way too costly, or the product, as designed, is just not able to be manufactured in the volume it is supposed to serve. Case in point, engineers need to understand the basics of manufacturing processes in order to design components in the most responsible manner. In other words, make the best part possible that can be manufactured at a reasonable cost.

GPS EDUCATION – TRAINING 'FROM THE INSIDE OUT'

While traveling in Wisconsin, I was introduced to another innovative educational concept. GPS Education Partners (GPS) is a concept like no other I have seen. It's just one additional way for us to solve the lack of a skilled workforce in the manufacturing sector in our country. While Cardinal Manufacturing mentioned in Chapter Seven, puts a business inside a school, GPS Education Partners puts a school inside a business.

The concept of Career and Technical Education (CTE), the current terminology in the United States for what has also been known as vocational or occupational education, has various meanings and forms of implementation. CTE instruction ranges from learning about career clusters, to taking course work in a particular vocational area, to some work experience that allows a student to "practice" what has been learned in classes. GPS Education Partners takes the CTE experience further and provides students with an immersive work based learning opportunity in which students graduate from secondary school with credentials that future employers value and for which they are willing to pay.

The purpose of work based learning is not a place where students who do not excel at academics go to acquire some training in the hopes that, one day, they will be able to get a job. The purpose of work based learning, as GPS delivers it, is to be a transformative experience for young people whether they plan on attending a four-year university, a technical school or immediately entering the workforce upon graduation. The opportunity to acquire and then practice soft skills (responsibility, teamwork, initiative, grit, etc.) along with specific technical skills is designed to prepare students for the current marketplace as well as the future in which they will be employed in roles that do not exist today. Each generation faces this same issue and must find ways to best prepare students for future careers. GPS is in the forefront of doing just that by providing deep apprenticeship experiences that go beyond career exploration to career preparation. This preparation results in graduates who are critical thinkers, equipped with skills and knowledge that are transferable to new situations and new technologies. By providing young people the opportunity to learn, to grow and to develop confidence while working in adult environments, GPS is assisting students in the transition to adulthood and a solid financial future.

The Organization for Economic Cooperation and Development (OECD) indicates that a strong CTE program includes a significant number of hours spent in a highly structured apprenticeship experience, coupled with associated classroom learning. For example, a student engaged in CTE in Switzerland will spend three eight-hour days per week, forty weeks per year in an apprenticeship experience and at least one other day per week in the classroom (OECD, 2010). This model is not commonly found in the United States, however, it is the model that GPS follows. Students in the GPS Education Partners program split their time during the week between working in their apprenticeship placements and attending classes for related instruction as well as traditional high school coursework (i.e. English, mathematics, science). Additionally, students are engaged year round in both apprenticeship and classroom activities to provide them with a continuous learning experience.

Students may enter the GPS Education Partners program as either juniors or seniors for a two or one year experience respectively. Strong relationships with the school communities and GPS translate into recruitment activities that bring students into the program. Once students are enrolled in GPS, they engage in traditional academic coursework as well technical material that, at first, is broadly related to the world of work and later in their experience, more narrowly focused on their specific apprenticeship area. For example, at the start of the program, all students focus on the soft skills around hireability, such as interviewing, timeliness, and communication. The parallel technical skills might be understanding safety in a manufacturing setting. Toward the end of their experience, their soft skill focus is on negotiating the terms around their first "real" job, while technical skills will be very closely related to the work they will be doing.

The close alignment of technical skill development with the apprenticeship placement is a result of the relationships that exist between GPS Education Partners and the business partners that provide apprenticeships. As the OECD report indicated, a robust apprenticeship includes highly-structured work experience. GPS works with each business partner to clearly map pathways that provide exposure to the various career areas in a business. These exposure opportunities provide students with a breadth of options available to them and from which they choose a focus area for career preparation. Clear pathways to career opportunities have also been mapped and lead to graduates who are poised to take on roles within organizations right out of high school. Without quality work plans, the apprenticeship can devolve into a source of cheap labor, resulting in no benefit to employer or student. In addition to academic knowledge and work experience, students acquire credentials, either through the work experience or through aligned technical education. GPS is committed to making sure that students graduate with options that make them attractive to business, but also leaves the door open to future education.

As the workforce ages, industry is unable to fill highly technical positions requiring less than a four-year degree; GPS Education Partners currently works with students, schools and businesses in Wisconsin to provide robust work-based learning for students that prepares them to step into those technical roles. GPS is the intermediary between education and business so that industry is able to meet the growing demands in the labor market by hiring skilled, credentialed employees.

A CONVERSATION WITH BRYAN ALBRECHT
PRESIDENT OF GATEWAY TECHNICAL COLLEGE

We heard from Bryan in Chapter Three. Gateway is currently in the mix of one of America's largest industrial development projects establishing Foxconn's $10 billion manufacturing facility.

Foxconn is one of the world's largest and most advanced producers of consumer electronics. Their product line includes Apple, Sharp, Dell and Smart Technologies. With over 88,000 patents, it is safe to say that Foxconn produced products are driving innovation in education, transportation, healthcare and entertainment. Designed on the platform of the internet of things, Foxconn is leading the world in smart technology innovation including autonomous vehicles, smart cities, Big Data and IIoT. It is the IIoT or the Industrial Internet of Things that has linked Gateway with Foxconn.

Founded in 1911, business partnerships are not new for Gateway, but technological advances in automated systems is transforming the way Gateway is preparing youth and adults for jobs and careers.

To address the skills gap and prepare for the over 13,000 advanced manufacturing workers needed by Foxconn, Gateway has established new programs built upon career pathways in automated manufacturing, Supply Chain Management, Data Analytics and Cyber Security. In 2018, there were 6,147 high school students enrolled in Gateway College courses earning college credit before graduating from high school. To enhance the talent pipeline, Gateway has developed customized training programs for corporate partners. Amazon, Kenall,

Snap-on, Wisconsin Oven, InSinkErator and Foxconn are just a few of over 160 corporate training partnerships that are operated by Gateway.

Local partnerships build global skill competence for students and faculty. Gateway is a strong influencer in the industry certification movement. With over 200 industry certifications, Gateway graduates are academically prepared, industry certified and job ready.

Working with industry is a driving factor in Gateway's investments in STEM education. Gateway's new Kids Lab Program is specifically designed for grades K-4. When young students, parents, teachers and counselors understand the importance of technology and engineering, their perspective of technical careers changes.

Southeast Wisconsin is stronger because of the commitment Gateway has made to K-12 education, university transfer opportunities and customized employer training. The National Association of Manufacturing, Manufacturing Institute and the Manufacturing Skill Standards Council have recognized Gateway as a leader in education and training.

With the advanced skill requirements that Foxconn and other industries are requiring, Gateway is strengthening their programming by integrating technical skills with cyber skills. This new technical training model will produce advanced technicians suited for Industry 4.0 and Smart Manufacturing careers.

I believe we are at the forefront of the next industrial revolution. One that includes artificial intelligence, augmented reality, additive manufacturing, collaborative robotics and virtual design. Each is raising the bar for colleges like Gateway to remain viable in workforce preparation. The careers of the future will require high performing technicians. Foxconn is expanding our vision and leading us into the future.

Terry, thank you for allowing me to be a part of your journey to help build CHAMPION Now!®

A CONVERSATION WITH MIKE READER
PRESIDENT PRECISION PLUS – WISCONSIN SWISS

For those who are not familiar with our industry, there is a term that describes Mike Reader's shop - Precision Plus. It comes from the watch industry when small very precise lathes were used to make components. These products have been manufactured in Switzerland for hundreds of years. Those shops that utilize these type of machines are called "Swiss shops." The area where Mike's company is located is well known for some of the best Swiss shops in the country. Mike's shop is certainly one of them, and he is a big advocate for young people joining our industry. He speaks to this daunting task for which he is leading the charge, both nationally and in Wisconsin.

I've been in this business for about twenty-two years. I didn't start out in manufacturing. I didn't grow up in manufacturing. When I came into the business, I couldn't understand why people weren't knocking on doors looking to learn about the careers in manufacturing. For a lot of years, I was on the sideline complaining about the lack of talent. What I really should have been recognizing is that the only way we attract and develop talent is through the engagement with the community, primarily the schools. How is a student or an educator supposed to be aware of the careers in manufacturing when we don't engage them? They're not gonna go out and find us. We've got to take the lead. We've gotta go out and engage them. We've been very focused at the area high schools, at local technical colleges, even the four year track with the Milwaukee School of Engineering. We are all ignorant of some things. If we're not exposed to them, how are we ever to connect the dots with students, to show them the beauty of what we do?

I've been at this, probably all in, for the last five or six years. That was only after having one of my high school friends along with acquaintances who have been in the teaching community for a lot of years, come through on a tour following my high school reunion. I could quickly see that he could understand what was going on in machine tools. I had a light bulb moment there that I've been trying to get machinists to build training curriculum for quite some time. They were never able to do it. I realized that I need to have somebody who is good with education building those systems. The challenge for us is trying to get everyone else on board. I was on the sidelines complaining for a lot of years. I can't be too hard on them. I was right alongside them for many years, complaining about the lack of talent, and nobody doing anything deliberate for us, rather than going out and engaging ourselves.

Manufacturing Day®, I think that's a great opportunity to open the doors. Invite the community in, show them the beauty of what we do, and only through those efforts will people go, "Wow, that looks pretty cool. How do I learn more about it? Do you have internships? Do you have youth apprenticeship programs? Can we do a job shadow? Can we do teacher externships?" Sometimes the education community is one of our biggest obstacles to overcome. They measure success by how many degrees that a person has or what percentage of the student population went to a four year university.

That doesn't mean they completed a four year, but they are measure with, "Hey, we put 90% of our kids in a four year university." Well, what they failed to look at is, what are the statistics? How many of those kids drop out after one, two, three years, and wind up with a pile of debt? Now, they're really lost. We've engaged a lot of the educators. We had to convert some of them.

Unless they've got an intimate knowledge of a forward thinking manufacturing group, the view of the parents is, "Don't let my child grow up and go into manufacturing. The dark, dirty, dangerous comes to mind: dead end. All of manufacturing is going to Asia. Why would you ever want to pursue a career in manufacturing?" It's almost like a second class citizen. Many are ignorant of some of the amazing things that go on in our industry every day. The parents are some of the biggest challenges. You can't get to the students without getting to the educators. You have to get the buy-in of the parents at the same time. That's why when we bring students through for either youth apprenticeship, YA programs, or summer interns, we require that they bring at least one parent with them for that first walk through. The parent can see that we are there to take good care of their children.

The craftsmanship (in Europe) seems to be appreciated and held in much higher value than what we have here. I've traveled through Germany and Switzerland. I've seen some amazing shops doing wonderful things. They're struggling at finding bodies as well. I think that might be just a demographic issue, as they've got a declining native population.

It's students, it's educators, it's parents, and then it's our elected officials who need to use the voice that they have on the state or the national level and highlight the importance of a strong manufacturing sector, if nothing more for national security reasons. We have had every elected official that you can probably name in the state of Wisconsin through our building trying to educate them. If you've not been exposed to it, you're certainly not going to advocate for it. We have a very strong effort trying to educate them at every opportunity.

I travel to Washington DC with the PMPA (Precision Machine Products Association). I met personally with Paul Ryan. He came through my plant to talk to us about what was the result of the tax legislation. I've been doing this for at least ten years with the political side of it. They now are starting to recognize, "Hey, this is an important voice. It needs to be heard in the manufacturing sector. We need to understand what's going on, what we do that helps or hurts them, and try to promote some of that. At the end of the day, we've gotta get the kids, the parents, and the educators through the building. We've gotta show them the skills that are required to do the things we do.

Bottom-line is, you have to get engaged. Other business leaders need to get engaged. That is the only path forward. The easy stuff continues to move overseas. The challenging stuff is what remains here. If you don't have the best and the brightest going forward to run some of these machine tools that are very sophisticated today, it's gonna be a tough road for you. You can choose to die by 1,000 cuts, or you can take the time to get involved with the schools and the community. Provide the career exploration options for the kids.

Automation and technology is creating new jobs. It is allowing people to work more with their minds, rather than just their bodies. These simple, repetitive tasks that can be automated, should be. Allow that person to grow and develop to do far more capable, engaging, (and) stimulating activities. Thank you Terry. I appreciate all that you have been doing for so many years in this area. It's important, and some days it probably doesn't feel very rewarding. Thank you for all you're doing.

We have covered many exciting things that are happening in both the states of Florida and Wisconsin. There are some very capable and visionary leaders, in Florida Marilyn Barger and Lake Ray, and Bryan Albrecht and Mike Reader in Wisconsin. The next chapter addresses that each young person needs to find a passion that matches their career objectives and life goals. Hopefully Finding America's Greatest Champion will educate, inform and inspire those looking for answers and opportunities. They are the Champions we are grooming. It is up to them to realize their potential and claim their CHAMPION status.

FIND YOUR PASSION, DESIGN YOUR LIFE

"Why did the eleventh one work? Well, I learned from those ten. I don't want to be known for ten failed companies but one good one. But those ten got me to the last one that worked."

– Craig Rabin, Founder of the AirHook

Our culture today often expects conformity, especially for young people, but each of you need to have the courage and confidence to do what's right for you. If you truly find something that excites you, it will energize you to excel and, in most cases, the money will follow.

A CONVERSATION WITH HARRY MOSER
LEARN ABOUT YOURSELF

One of the people who inspired me to get into the education path is the same person who founded the founder of the Reshoring Initiative ® . I became inspired by Harry Moser, by just watching the things he believed in and what he did. His work has influenced many in the boardroom and C suite offices to turn the spreadsheets inside out and consider the full cost of ownership when making decisions on where to make your goods. Here, he shares commentary on choosing a career and balancing this with your passion. Around 2010, after retiring from the machine tool industry, Harry founded a movement that became known as RESHORING. Both Harry's passion to inform students about manufacturing careers and his reshoring efforts were more inspiring than anyone else I had come to know. Two tremendous efforts and well placed passion. Harry Moser is someone that I respect and admire. Here are some of his comments:

As a country, we need to give young people encouragement and permission to find their passion. Explore how young people can do this effectively. Take a look at skills assessments, the high schools who are doing this well, why some guidance counselors are doing this well and some aren't, and other supporting information. How do you know if you are a good fit for manufacturing? How about listing the traits manufacturers look for in people (respect, curiosity, hunger to learn, diligence, problem-solving, etc.) and the strengths one should sharpen (goal setting, listening . . .).

Get a career that will pay for you and your family. Develop a passion for that career. Pursue your other passions using the time and money generated by your career. Work entitles you to what you earn and desire.

I want you to meet young people who are fascinating manufacturers/ entrepreneurs and who understand their journey. These people went against the grain of what culture tells them to do. And they won!

NFL & HOLLYWOOD – PROMINENT IN THE AMERICAN CULTURE, BUT WHAT ABOUT THE MAKERS?

In 1985, Kathy and I moved to Illinois with our (then) two children. (We had our third in 1989.) We chose to move to Long Grove, Illinois and send our children to one of the best public schools in the state—Stevenson High School in Lincolnshire, Illinois. My daughter Lindsay's class (graduating 2001) in high school had some pretty interesting success stories. I am sure that other people at Stevenson made it, but the stories that I am going to choose to tell are the ones that are personal to me. Back in the late nineties, Stevenson was known for some famous alum like the international sensation tennis player Andrea Jaeger, the Ryan brothers in the NFL/ESPN (twin sons of Chicago Bear Coach Buddy Ryan—Rex and Rob), and a few other NFL players among others.

While our culture looks at Hollywood and professional athletes as iconic people to look up to, I would like you to consider, "What if we looked up to both entrepreneurs and manufacturers that make our country both prosperous and successful?" Taking nothing away from any of these success stories, I just would like manufacturing stories to get the attention they deserve.

AMERICA'S GREATEST CHAMPION

What is America's next greatest Champion? Who is America's next greatest Champion?

There are so many definitions. You could be it. Manufacturing could be it. The female gender could be it. There are so many possibilities, and none are incorrect. In this book, I plan to tell many stories, some of which I hope inspire young people in this country: parents, influencers of our youth, educators, entrepreneurs (current and future) just to name a few. Having said all that, I would like to talk about some young people I have met when they were quite young, who have inspired me with their vision and the direction in their careers.

One young man who I got to know from Lindsay's class is Craig Rabin, a modern day inventor and entrepreneur who took a chance, followed his passion and successfully brought a product called the Air Hook to market. The Air Hook is a product that you take on a plane with you that allows you to hang your coat, mount your electronic device for viewing, or hold a drink—all while your tray is in the upright position.

Craig was pretty much a techy geek from day one. He is a very gentle soul and genuine to the core. During the early days of internet and web design, he was very entrepreneur driven, and I used to pay him hourly to do work on my iversonandco.com website. He would come over to the house, and the two of us would work on my website. We would huddle around my laptop while Craig did html code to develop the website that would expand Iverson & Company's presence on the web. Craig helped me on my second revision of the site around 1999-2000. Eventually Craig would get hired on at Microsoft and move out to Washington. Like a lot of young people who I try to keep in touch with, we kept in touch on Facebook.

Craig was fortunate enough to be one of the very first contestants on Steve Harvey's show "THE FUNDERDOME," a business reality competition. On the show, entrepreneurs compete head to head for money to advance their products to the next level. The audience then votes on which of the two should get the money. Craig won the first round and the twenty thousand dollar prize, making him also the very first winner on the show.

A CONVERSATION WITH CRAIG RABIN
"THE UNEXPECTED INVENTOR"

From a young age, my dream was to be an entrepreneur. The real goal was to be an inventor. But I wasn't a super book-smart person. To be honest, my brother who is five years older is a frickin' genius. So coming up after him, it was a lot of pressure to be smart. That forced me to be street smart, always coming up with different things to fill that void. I was in the library for hours.

Years went by. I was thirty; I was single and experienced a third-of-a-life crisis. I had a great corporate job at Microsoft. Do I try my hand at being an entrepreneur? I decided I wanted to take one more chance. I had this notebook of ideas of different products and concepts. My pivot in the road was 3D printing, so I could get a machine in my house that could do the prototyping. So I bought one, and for a year, taught myself how to use it. I thought of Air Hook when I was sitting on an airplane and had no idea where to hang my sport coat. Inventing it was amazing. I had to relearn geometry.

The only piece that didn't air on Funderdome was the cancer ribbon that I wear. That was because we do a donation on every Air Hook in honor of my mother who passed away from cancer in 2016. We want to be that leader. Even the smallest brands and companies can help change the world. Cancer is something we all have to fight together. I want to continue to fight for other people. It's been a really rewarding part of this journey. Every sale represents one step closer to a cure for cancer. It boils down to that. It gives me and my team a push forward. Every day, we're accomplishing something for a greater good.

It was hard for me to figure out what I wanted to be. All I knew was I liked the idea of owning a business in developing markets. My guidance counselor had absolutely no clue what to tell me! For me, what has worked when there is an unknown is to ask for help. I remember thinking how hard it was to figure out what I wanted to be. I didn't fit that norm. I didn't need a guidance counselor telling me—who didn't know what it was that I was talking about. It wasn't until I started asking people I looked up to, how they got there and what advice they had for me.

In some stories, the young people are quick to admit that they are not the smartest sibling in their family. These young people were/are bright and talented, but, equally as important, they are dedicated, motivated and driven. Young people need to toss away inhibitions of what they might consider limitations. This holds them back. Instead focus on your strengths, and get the most out of them. Knowing your limitations and understanding how to compensate for them is important. Your confidence and persona going forward depends on you knowing who you are and what your passion is. This gives you the recipe for success. I was so VERY fortunate to have people mentor me. Not only my dad, but coaches, teachers, and friends of my mom's.

MY EDUCATIONAL STORY – AN UP AND DOWN SCENARIO

I was born in Chicago, the manufacturing mecca of the USA, during the late 1950s. My parents ended up getting divorced, and my sister Kelly and I moved to Jacksonville, Florida. Jacksonville is a great city, but not (yet) exactly a manufacturing hub. I was always a smart student, excelling in math and science. I started in public school, however my parents thought that private school was a better match for me. I took the entrance exam for The Bolles School. Somehow, I got in! Bolles was previously an all boys military academy then became coed in 1972, the year I was accepted. Bolles is a school that I believe set me up for success at a young age. I was able to balance academic expectations with social, athletic and other priorities (work). Overall, I came out with a great high school experience. I hope the same for all young people. This is why it is so important to present all opportunities to our youth.

Even today, I feel attending Bolles was the biggest break for me both personally and professionally. The environment I was in was the most challenging I had ever been in. Excellence and achievement was encouraged, even expected. It was here that I learned the art of over achievement—performing over your natural abilities through persistence and hard work.

Throughout most of my education, I was bored; I wasn't challenged. There was way too much theory. I always made sure I enjoyed everything I did: the social scene, athletics, and, for a while, academics. Once I got into AP Physics and AP Calculus, there seemed to be way too much theory.

Outside of school, I worked most of my junior and senior years. I played football and soccer and excelled in the latter, as the leading scorer on the team. Most nights that I worked, I got home about 1:00 a.m. and had to be up at 6:30 a.m. for school. My AP Physics teacher (nicknamed Boo Boo) told me one day that I would be academically ineligible for Friday's soccer match, as I had scored a wonderful "50" on my test. (Needless to say, I had worked a lot that week before.) Now, he had my attention!

I would not/could not let my soccer teammates down. I asked my teacher when our next test was, and he said "Thursday." "What would I have to make on the test to be eligible?" I asked. I added, "Would you grade it that afternoon to see if I could play?" He agreed and said I would have to make a ninety-five in order to play. I stated "Ok then. Deal." Boo Boo was stunned at my confidence and matter of fact statement.

"You really think you can do that?" he asked. "I know I can. It's all about priorities. I will have to take some shifts off at work, but I WILL get the grade." That Friday, Boo Boo gave me my test back—on the top of the test was the grade "97." He shook his head in disbelief. ("Why couldn't he do this every test?" I am sure he thought.)

Friday's game I started at center forward —disaster averted. Boo Boo challenged me. Much as he didn't understand, he made it real for me. He was awesome. He cared.

COACHING AND MENTORING FUTURE AWESOME ENTREPRENEURS

As my wife and I started a family, I kept soccer close to my heart. As our children grew, I felt this was not only a great sport for our children, but also a chance to stay connected with them as they got older. Years later, when I was coaching traveling soccer, I helped coach my daughter's team. They had a really good team with some awesome athletes. Our daughter, Lindsay, was faster and tougher than anyone could believe. One year, they lost in the Illinois State Cup final with a couple key players ill, or unable to play. One of the young ladies who played for us was Dana Ward. She was a good athlete, maybe not the best soccer player on the team, but a big heart and very athletically talented. She also was probably the most well-rounded person on the team. Not only did she play at a very high level on the soccer team, but was also involved at high school on the pom squad, while also doing modeling. Dana was the closest young person who, similar to my story, made a conscious decision to have a well-rounded high school experience. She went to a very hard and demanding high school and also managed to achieve good grades. Dana had it all going on at a young age.

A CONVERSATION WITH DANA WARD
"THE IDEA MAKER"

The first time I came across Dana out in the working world was when I was searching online to read up on one of the Chicago Bears draft picks. Dana was on an online TV show that covered upcoming college football games and who was injured for an upcoming game, etc. I called out to my wife Kathy, "Hey I think I see Dana online. Come here and watch." From there, Dana went into ClevverTV where she would be on the red carpet interviewing Hollywood stars. Just like some of the others I am writing about, she then crossed over to the entrepreneur/manufacturing side when she developed a new product called Pre-Heels. This is a product that can prevent women from getting blisters from new shoes. Here's Dana's story.

From a young age, I've always been plagued with noticing little details that others don't seem to see, and it's been important to me as an entrepreneur to not just improve existing ideas, but to actually develop innovative concepts and products that don't already exist. The world doesn't always need another (fill-in-the-blank) website, product or company.

The idea of innovation was something very important to my cofounder and me ahead of our PreHeels product launch. We created the blister prevention spray from scratch over an initial three years of research and development before our Fall 2016 launch. Without getting into too much detail, saying the process was a difficult feat is an understatement, but continued development on this and future formulations as we expand our line, is essential to our company and what we stand for. For our initial product launch, we chose to 'soft launch' via a popular crowdfunding platform to validate interest, test the audience, garner reactions, get customer feedback, etc.—an important survey step in any entrepreneurial strategy in whatever format or platform you choose. If you listen, they will tell you. We went on to pass $3 million in sales our first year, and we're continuing to optimize and innovate as we work hard to grow!

As you might be able to see, I'm a pretty optimistic person, and I think that's important—at least to me—in entrepreneurship. Sometimes it feels that if it can go wrong, it will go wrong, so you need to be able to keep your head down and keep working through the storm in hopes that the sun eventually comes out to shine. Now I don't want to scare people away from entrepreneurship, but I think it's important to understand why you're choosing to launch something. Don't do it just because it is trendy or the cool thing to do. Entrepreneurship is not for everyone. It can be extremely difficult on your mind, your body, and your soul. It requires sacrifice, self-motivation and calculated risks. Just be honest with yourself about why you're building a product or business, and ask yourself how it is new or better from what already exists in the market.

I'm very happy that you're writing this book. I think we all need a refresher on this topic and seeing the entire picture and multiple examples is always helpful; there's so much noise out there with digital media, so it's too easy for us to be moved by a thought or action-item for a split-second but then unconsciously move on, swipe up, double-tap something else in the social media feed, and the inspiration is gone. Let's keep the stories and framework in a single place for easy access and promising follow-through.

A CONVERSATION WITH JOHN SAUNDERS
NYC CNC YOUTUBE VIDEO KING

One such person who did not know manufacturing, but wanted to be an entrepreneur, ended up teaching himself to BE a manufacturer. In doing so, he decided to give back and post YouTube videos on his manufacturing lessons. His name is John Saunders, and he goes by the YouTube moniker NYC CNC. He has 229,000 subscribers on YouTube.

I grew up in Ohio where I now live. I had my sort of initial exposure to manufacturing when I was working on our family farm, and my grandfather was a fabricator. He had retired by the time that I was hanging around with him. We had a welding machine, and we did various sort of repairs on the farm. Fast forward, I went to college for entrepreneurship. I still had a passion for this idea of making and being hands-on. I didn't know anything about how to get something made. I knew nothing about manufacturing and machining.

I was living in New York City working a day job totally unrelated to this. I had this passion and determination. I bought a benchtop CNC machine. This would have been late 2007. I mention the timing because it's sort of relevant to what's been happening in the U.S. In America, there has been this really awesome renaissance of not only an interest in manufacturing, but there's reshoring. There's been maker spaces and hackerspaces and things like Arduino and 3D Printers.

I bought this benchtop CNC machine. I'm trying to figure it out, and my real goal was only to be able to talk to a machinist. I didn't really think I'd ever become a machinist or a maker. I didn't think I had the ability to. I thought you had to go through formal training and apprenticeships or have a family machine shop to really learn that. But I totally fell in love with it. The idea of being able to take an idea, CAD (Computer Aided Design) it up, CAM (Computer Aided Manufacturing) it up, use tools, use raw materials to make that part, to me, is just such an awesome thing. I really love that.

When I was doing this, I realized I had a digital camera, and I enjoyed doing stuff with that before digital cameras became ubiquitous. In 2007, people didn't know what YouTube was. I just got lucky. A friend had mentioned it to me as a good new online video site, which is funny to hear someone having to describe what YouTube is nowadays. I thought, "You know what? If I'm interested in this and I know nothing, there's got to be at least ten other people who are interested."

I wanted to have some goodwill built up. When I learn, when I get help, I intended to try to pay it forward and share it with this YouTube Channel. Fast forward, I started to learn more about machining. I kept trying to figure out "Hey, where am I going to fail? Where am I not going to be able to figure something out?" I just kept figuring it out.

I thought well, what do I want to do? I really love this. I thought well, the YouTube Channel has certainly gained some popularity. It was still pretty small at the time: maybe 15,000 or 20,000 subscribers. I thought this is something I enjoy. I know I'm doing a good thing. I know we're inspiring people. What bothered me was, I wasn't upset that I didn't learn about machining until later in life. I feel if there's a thirteen-year-old or a fifteen-year-old, it's okay if they don't like machining, but I don't want them to not see what it is and what the opportunities are. Today everyone must realize that machining is not working in a dark, dingy, dangerous factory.

I have a little bit of a different view on this sort of thing. I will do everything in my efforts to provide the opportunities and the resources to anybody of any age, especially young people, who have a passion and have a desire, but they have to have it. They have to have that spark. I had to work hard. If you're motivated, if you're driven, if you've figured something out, then great. Let's make sure you have the resources. Let's make sure you're not limited by money, by opportunity, by risk, by access to any chance to grow. Let that passion flesh itself out..

I'm a really big believer in avoiding debt. It's different when we're talking about capital financing for large corporations and so forth. At the individual level and at the small business level, I really discourage folks from getting in over their heads, borrowing money and living the lifestyle they can't justify and afford.

PATENT PENDING GIRLPOWER!

I served on the Technology & Manufacturing Association (TMA) Education Foundation committee for a few years. Members on the committee suggested that we take our meetings out to the schools and organizations that we approve grants for. One such meeting was at Northern Illinois University. There was a grant submitted for a group that called themselves the Moovers. After reading over the grant, I became aware that this was a group of young women who competed in Lego Robotic competitions against primarily male competitors. "How cool is that?" I thought. We carefully read their submission for funds to support their activities. It turned out they had accomplished quite a number of successes and awards. Further reading lead me to notice that they were located in the suburb of Chicago that I lived in, and they attended the school our youngest son attended. This made me even more curious. I took interest in their vision and goals and told myself that after they were notified of the grant, I would become more active in their activities.

Soon after the grant was awarded by the TMA, I gave their mentor leader a call to introduce myself. I called and stated that I was a member of the TMA Education Foundation and that I was interested in their group. They thanked me and the foundation for the grant and invited me out to one of their weekly Sunday night meetings. We scheduled a visit. I got to introduce myself and told them how fascinated I was by their group that by all descriptions was an anomaly for their gender and age.

We talked and traded questions and comments while they indicated interest in knowing and understanding more about manufacturing. Unfortunately, I was not able to visit them in August before the biannual International Manufacturing Technology Show. I was disappointed to have missed a great opportunity to show them a fantastic demonstration of the excitement of manufacturing right in downtown Chicago. "Two more years," I told them, "Besides, all of you are so young. You will only be fourteen at that point!" I left that night with a new sharpened vision on what young women in this country can do to make a difference. The Moovers certainly could move our workforce by example into different times where women could be a much more significant percentage of the manufacturing population in the U.S.

A CONVERSATION WITH STEVENSON HIGH SCHOOL GRADUATE ANISHA RAO

As years progressed, I kept in touch with one of the young Moovers, Anisha Rao, through her mom. I once met with their group with Congressman Brad Schneider for a photo op. In 2018, Anisha, finished her senior year in high school, and embarked on her collegiate journey at University of Illinois, with a focus on engineering. She interned with Iverson during the summer of 2017, and she has a great future ahead of her. Here is some of Anisha's story including her project that has great promise:

Girl scouts, with help from a grant from FLL—which is the First LEGO League for robotics—allowed a couple of groups to start robotics teams. They were all girls, so that's how I got involved. My mom really encouraged me. Ever since I was little, I would build devices and try to create as many things as I could. She figured that I might as well use these skills in purposeful fashion.

LEGO robotics is a program for ages nine through fourteen. There are different portions of the competition. One is called "The Game." Every year, they release a new kind of board. On this four foot by eight foot board, there are different tasks. Each team has to build a robot out of Legos in order to compete.

They try and complete as many of these tasks as they can in two and a half minutes, and get as many points as they can. The other portions are called the projects, so every year there is a team to build the game and the projects.

We had qualified for a national competition through this program, and so we were looking for sponsors and people who were able to mentor us. We went around the Chicago area and the TMA (Education) foundation was one that came up multiple times.

At Stevenson High School, we have Project Lead The Way, a national program that has actually partnered with the Advanced Placement Program. They provide different courses, ie Basic engineering, like CAD work, and electrical engineering.

The advice I would give young girls is to not let other people deter them. People may tell them, "You're not creative enough or you're not problem solving enough, you're not good enough at math, you're not good enough in science." They shouldn't let that stand in their way if they are really passionate about what they want to do. They should pursue it. Do not let the lack of women in the field also get in the way. The only way to fix that issue is if more women go into the field. There is a lot more support for women in engineering now. They should really take that into account.

The older we get, the more responsibility we have to help the younger kids realize that they're capable of this. If WE don't do it, who's going to do it? With an all-girls' team, especially starting at a young age, it gets rid of any deterrent from other people. It was because of no one else telling me that I couldn't do it—that I was able to do it. It was an all-girls' team. When I made a mistake, I felt like it was because I didn't understand—not just because I was a girl.

My goal one day is to be a CEO of my own company. Kids have to start realizing from an earlier age that girls are able to do exactly what boys can do. The overbearing amount of work that students have to endure here is not conducive. There needs to be more room for kids to pursue a passion of theirs rather than loading up on another AP math. Can't they take an elective like an engineering course, or take an elective in arts? It should be less about building a college profile and just allowing students to learn and enjoy learning, and enjoy their experience in high school, rather than having to worry and have four hours of sleep a night.

We have to focus on the women right now. We have to focus on the younger girls, especially still in school, in order to fix the gender gap that we have. The young girls should not bring each other down, but also help each other rise up. They'll bring you up, all of us together and not just yourself.

PATENT PENDING ...... WHAT LIES AHEAD?

Over the past two years, I have developed an on-site diagnostic test that can chemically detect concussions. According to Dr. Evan Scott, a biomedical professor at Northwestern University, once the provisional patent comes through, I am likely to gain funding to start a clinical trial. Within three years, I hope to make my product available to thousands of sports programs in order to reduce the number of undiscovered concussions among youth athletes.

The current concussion survey is administered by officials to determine who is eligible to continue playing; however, it relies upon players to provide honest responses. When gamed by a competitive and ignorant teenager, the survey is useless. I know this because on three separate occasions, I was that teenager. Eventually, I was told I was done playing soccer.

When I found that I could no longer play, I was disappointed but more so, I was angry.

I take full responsibility for my actions. How could someone in my place be expected to take herself out of the game when she had been taught to always "take one for the team." A twisted ankle? Keep going. A sprained wrist? Shake it off. An elbow to the head? Voluntarily take a seat? There needed to be something to keep us from ourselves. I embarked on a journey to discover what that could be.

I began by researching what a concussion was. I scoured the internet to understand how they are categorized, how they are diagnosed, and how they are prevented. I was shocked to discover that the on-site diagnostic system had not progressed past what I had experienced. Why? I delved deeper into diagnostic tests and studies to answer this question. There reached a point where 75% of what I was reading consisted of terms and systems that I had never heard before. At the time, I was taking AP Biology, so I reached out to my teacher to help me understand them. When the specialized articles had surpassed his level of expertise, I began relying upon the smartest teacher I know: Google. I spent hours upon hours searching for definitions of specific proteins and how complex biological systems worked. Finally, after months of research, I encountered an article suggesting a viable, biologically-based way to diagnose concussions. I took this article and ran with it for the next two years.

Research. Design. Prototype. Test. Analyze. Repeat. This became my life during junior

year. I started with rough sketches of potential prototypes that could chemically test for concussions to try to visualize the ideal thickness of the casing, the type of ANSI standard I would use for the threading, the radius of the chamber, et cetera. My sketches became increasingly complex, with haphazard arrows and crossed out annotations everywhere. I would then render them in 3D on CAD software, usually Autodesk Inventor. After creating a prototype, I would 3D print them and test them for any structural faults or design imperfections. After discovering flaws, I would head back to the drawing board and restart the process, refining, improving, and simplifying my design over twenty iterations. I spent my summer between junior and senior year in contact with my mentors, Dr. Evan Scott and Mr. Iverson, my boss from my manufacturing internship, presenting my progress and then heading back to the drawing board to work on their critiques.

Though I am a few legal steps away from human testing, I believe that in the near future, my product will be able to more accurately diagnose concussions in sports. The patent process is long and arduous, but I am patient (and unapologetically ambitious). If I only earn a bachelor's degree at the end of four years, I would be unsatisfied with myself. I hope to not only earn an education, but to also put what I learn to use in developing a viable product that is market-ready.

MY PERSONAL JOURNEY WITH COLLEGE MAJORS AND CAREER CHOICES

When I was deciding what to study in college, I was naïve as to what I wanted to do as a career and what I wanted to study. I had a newfound interest in photography. My mom had given me a nice camera for high school graduation. Upon entering general studies at Florida State University, I thought both photography and psychology held promise as far as potential choices for a major and subsequent career.

After careful thought, I came to the realizations that even though sports photography was an awesome dream job, how many photographers would be on the sidelines for any variety of sports? I came to understand that not only is there a limited number of opportunities for this position, but also the pay was probably somewhat limiting.

When I thought about psychology as a major, I realized that in order to succeed in that field, most people went on to grad school. Despite the fact that I thought it was a fascinating subject matter, the thought of sitting and listening to people's problems all day long, quickly deterred me from pursuing that major.

And then came my strong suit – math and science. I decided that if I was to go into engineering, that Florida State University was probably not the strongest choice at the time for that subject. I then decided to transfer to UW Madison in Wisconsin. I still enjoyed taking photos of sporting events and continued this well into my education at University of Wisconsin Madison. There, I became a staff photographer for the school newspaper. I was now at a strong engineering school but determined that I could still satisfy my interests in photography in a way that was not a career choice.

My college career took many turns. I transferred several more times as I struggled to find the match for me. This was more complicated with the fact that my girlfriend, at the time, was still in Florida. Long distance relationships are very difficult to manage—to put it mildly. Pursuing our relationship led me to transfer back to University of Florida Gainesville, get engaged and move back north to Wisconsin. There I took a job with my uncle's machine shop and went to Marquette University for engineering classes in the evening.

By this time, I had decided that my goals for earning expectations and my strong math and science capabilities were a great match for one of the family businesses in manufacturing. Even though I did not finish mechanical engineering school, the background that I accumulated along the way was more than satisfactory for the machining or machine tool business that I ultimately ended up entering into.

A CONVERSATION WITH FRANK FRANGIE
A SIDE STORY ABOUT FOLLOWING YOUR HEART

I met my future wife-to-be in my junior year in high school, while working at a movie theatre. One of my co-workers came to a movie with one of her good friends. The friend (Kathy) was beautiful with long hair. When she came in, she wore a neck brace. I thought, "Wow, this girl must have been in a terrible accident." Being a "compassionate person," I thought I should ask how she got hurt. She went on to tell me that she had pulled a muscle in her neck while putting on a jersey. I laughed out loud. (Isn't this where I should have said, "Did I say that out loud?") Not exactly the best first impression eh? What a fool I was back then. Anyway, I called her on the phone after. For some reason, she didn't hang up on me!

As we became serious and close to engagement, Kathy indicated there was one person I had to meet and get "his approval"—her best friend in high school. You guessed it—Frank Frangie. Fortunately, he approved, and thirty eight years later we are fortunate to have a very blessed marriage and life together. A better partner, supporter and spouse I could not have found. As the following remarks by Frank reveal, following your heart applies to career paths in much the same way it does to finding a life partner. In both cases, you need to trust your gut instinct, risk your heart fearlessly and commit 100%.

FRANK FRANGIE - WHEN YOU ARE A CHILD . . .

I grew up wanting to be a sports writer. That's all I ever wanted to do. I graduated from college in 1980. There weren't a lot of sports writing jobs open. For family reasons, I took a job at a bank. I was in the personnel and the human resources department. I learned a lot about how to lead, managing people, teaching leaders to lead, and how to develop young people, all of the stuff that I had no idea that I didn't know. Sometimes you don't know what you don't know, right?

Even though I was full-time at a bank, I worked part-time as a sports writer. I did that for two and a half years, then I realized I really wanted to be in sports. And I said, "Okay, that's not me. I'm a sports writer." So I came back into sports writing in the summer of 1985.

I knew very early on, at that bank, that that wasn't me. I was gonna be a sports writer, and it wasn't ever about the money. That was never the plan. Hell, I grew up without any money. We lived in a 900 sf home. It was never about the money. I just realized that the banking thing wasn't for me. I got out of that pretty quickly. The most interesting career change was the one from sports writing to broadcasting. That was the one I never saw coming.

I was a sports writer for a while. Three years later, I got a chance to do radio. I loved it. I worked both jobs, kind of a sports writer by day, radio host by night. I went right back and went to work with Jay Solomon in radio and continued to do both sports writing and sports radio until 1993, when I went into radio full time.

I tell young people, "Number one, before anything else, find your passion." If you find something you love doing, you'll never work a day in your life. Wake up in the morning passionate. Go to bed at night passionate. Drive around during the day passionate. Think about it at lunch passionate. Make yourself not think about it at church, that you're so passionate about it. Find something you're that passionate about.

After talking to Frank, I wondered to myself what would have happened if he had stayed at the bank. I think he would never have had the stories he shared with me—his eyes on fire thinking about the exciting experiences he had. Looking in my rear view mirror, I can honestly say that I followed my passion too. So much so that it led me to writing this book and the opportunity to share my passion for youth, manufacturing and readers like you who care about those things too.

THE EARLY COLLEGE MOVEMENT – ONCE YOU KNOW YOUR DREAM

In Jeffrey Selingo's book College (Un)bound [1] , he speaks to colleges that prepare students in any number of ways. In his chapter, "Future Forward" he talks about one college in North Carolina that gives students an early start on a degree.

Each year, Wake Early College of Health and Sciences in Raleigh, North Carolina selects seventy-five eighth graders from the greater Raleigh area and puts them through an accelerated program that in four years (or sometimes five) gives them a high school diploma plus a community college degree, diploma, or certificate in the health sciences field.

The students attend classes at Wake Technical Community College along with older students of college-age. When they graduate, roughly 85% of Wake Early students enter four-year colleges with credit for two years of community college. The other 15% or so remain at the community college and with a few more courses, become certified for such occupations as nursing assistants, emergency medical technicians, and phlebotomists.

Wake Early is one of 70 similar programs in North Carolina which is the leader in the early college movement. Programs are also underway in California, New York, and Texas. Early college might not be right for every thirteen or fourteen-year-old.

Some have to sample college before they can decide on a course of study. But for determined teenagers, especially those with parents struggling with finances, this fast track might be just the right speed.

This is an example of a different path of study for those students who, early on, know what they want to do for the rest of their life as a career.

Earlier in this chapter, we heard from some young entrepreneurs and makers. Here we get to hear what they have to say about the importance of finding your passion.

CRAIG RABIN – "ASK FOR HELP"

I talk to high school students all the time, and I tell them, "When you find that thing you love and you don't look at the clock, it's fantastic."

It's always harder to ask for help than it is to give help. Failure is good as long as you learn from it. If you can turn your failure into learning, you can turn it into success. I've had eleven companies. Why did the eleventh one work? Well, I learned from those ten. I don't want to be known for ten failed companies, but one good one. But those ten got me to the last one that worked.

A CONVERSATION WITH DANA WARD
"BE AN INDIVIDUAL"

It can be difficult to embrace all of the characteristics that make you unique. I know that growing up, there was so much homogeny to fit in, be cool, etc., and I really tried hard to always be true to myself. I think it's a good thing that individualism is much more celebrated today than it was when I was young (almost to a fault, actually, but let's focus on the good right now). Think about it this way: if you're so used to trying to fit in, passively going along for the ride or even just pursuing the path that someone said you're supposed to take without thinking for yourself for a moment, you might be missing out on a class that's more interesting, an internship that's more inspiring, a project that's more lucrative or a career that's more personalized to you.

And it's OK to not know what you want to study or what you want to do for a living (hey, I'm still evolving!), but make sure that you actually take a few moments to think about it before you set off on some path because society tells you it's a good one. Think about your strengths and passions and consider future employment trends. Consider YOU. Be an individual. Think for yourself. And if you're trying to help someone else find their career path in this wacky world—whether a parent, counselor, writer, mentor or friend—perhaps you see something in that person that they don't see themselves, so share it. Remember that the re-education process from individuals and parents to mentors and teachers to society and media takes way too long and is on an extreme delay, so think for yourself. Do your own research, and find the best strategy for you.

Above Dana says you need to do your own research. Additionally, many young people get exposed to various subjects and concepts during their high school experience. Many times, this leads to them finding their passion for their careers.

A CONVERSATION WITH CARRIE J. KURCZYNSKI
"HANDS ON LEARNING"

Carrie Kurczynski grew up in our church that her parents attended. We will hear more from Carrie in Chapter Eleven. Here she cites examples of how she found her way into engineering and manufacturing.

I was lucky to have hands-on teachers who taught us to "see" the theory, not just memorize it. In high school, I remember my physics teacher laid on a table with a board across her stomach and challenged students to hit the board with a bat to teach the dispersion of energy. When I was in college, again our physics teacher dropped a stuffed gorilla and marble from the ceiling of the lecture hall at the same time to demonstrate terminal velocity. GE saw my experience working in the machine shop at Purdue. They saw that my senior project was being a part of the Mini Baja team and my hobbies included working on motorcycles. They directed me to apply for a field engineering position, where I ended up spending the first five years at customer sites leading teams in tearing down and inspecting steam turbines.

A CONVERSATION WITH TONY SCHUMACHER
"OVER PREPARE AND THEN GO WITH THE FLOW"

In Chapter Two, Tony Schumacher talks about making products in America and the choice we make by purchasing products outside this country. Here Tony talks about so many of the young people who he has given speeches to and how each young person should make a conscious decision of how to advance their young lives to the next level as they near graduation.

I did eight speeches the other day for eighth graders. Most of my speeches started with asking, "How many have had a test this week?" Every hand will go up because they're in school. I'll say, "How many of you studied so much that you were going to get an A, and not squeak by and get an A, not cheat through an A, not guess at stuff and get an A—actually understand the material because what the teacher is trying to do isn't give you an A or a B or a C, it's to teach you how to learn a subject. When you go off and get a job, you can learn it. The school's going to teach you how to learn."

I just ask them at a young age, "Why not become the person today that you want to be?" None of those kids look in the mirror and hope to be decent their whole lives. They all have big dreams. They just forget to study. Those dreams become more difficult to rise up to when they don't prepare. You study for a test. You try hard. You learn. You dive in. What I've learned with the Army car is the word adaptability. That's a beautiful word. If you're educated, you can adapt. You can make the changes in your career path because you're able to learn. You're able to dig in and go for whatever the path takes you to.

I think it's important that kids reach out. Now the degrees are not nearly as important. They're not nearly as important as being able to understand, learn, and jump in with both feet and say, "I'm part of a team."

Your best advantage as a youngster is to say, "I want to think outside the box." They're educated in a different way. They spend more time on YouTube than anyone I've ever seen. My son, Anthony, is a great example. He spends endless time, when I argue with him about something, he researches it. He goes and he researches and he comes back with valid points. As a young person, trying to get to show the boss that, "Look, I didn't walk in here to get the job not knowing what it entails. I studied your company. I know what your game plan is. I know what your marketing plan is. I understand that I'm just one person helping out here. I show up for work on time, and I stay late," is so important because it shows you're trying to learn. You will to be there longer because you have input. You have an interest in the company.

I think my favorite expression that I use is just to over-prepare, and then go with the flow. Too many people show up and just wing it. If you showed up for a test, just winging it, you wouldn't be very good at it. I think what that entails is number one—choose your job. Don't let it choose you. Choose it, figure out what you want to do. If you won the lottery, what would make you wake up and be happy every day? Then, you learn it. Don't learn it like you're trying to get an A, learn it like you've got two brakes on a car with two car seats in the back seat, and it has to work. Learn it that well. It has to be done right. If you can do your job like that, you'll be proud. Everyone around you will be proud, and let me tell you something man, when you do a good job, your friends want to do a good job, and so does everyone around you. It starts the world in a better place. I just think we're in a little bit opposite position right now. We need to fix it.

A CONVERSATION WITH JOHN SAUNDERS NYC CNC
"YOU'VE GOT TO HAVE A PASSION"

Earlier we heard from YouTube machining guru, John Saunders. Here he gives legitimate commentary on the balance between finding your passion and also being able to balance lifestyle expectations with career choices.

People need to be comfortable and okay and recognize that most people in life want to earn some amount of money and also be happy. I think it's incredibly rewarding and, again, right now, knock on wood, we're in a pretty darn good economic environment where there's a pretty huge demand for skilled labor and for trained work. Again, you've got to apply yourself. You've got to do some work. You can't be lazy. You've got to have a passion. You've got to stay away from drugs. You've got to be reliable. You've got to be able to read and write. I almost am embarrassed that I have to say that stuff.

Two of my favorite quotes both encapsulate the same concept. I'll give you the gist of them here.

I don't care what you do as your career in life and vocation, but recognize that it needs to be something that can justify and sustain the lifestyle that you want to live. If you want to do something that may not offer you much opportunity to earn a lot of money, that's okay, but just recognize that's going to have some limitations. The other quote, which is great, is from Warren Buffet when he's actually talking about his children and their legacy and his inheritance, and his comment essentially is, "I want to make sure that they have enough that they can do anything, but not so much that they can do nothing."

WE NEED THE BEST AND BRIGHTEST – THAT'S YOU!

In this chapter, you have read about some fascinating people. Each of them has an interesting story on how they discovered and nurtured the Greatest

Champion inside each of them. But what about the plethora of young people who may not have the support or intuitiveness to find their way—especially in the field of manufacturing?

Having said all of this, the next generation holds the key to what we will ultimately be able to accomplish in this country. Manufacturing can be brought back into the country, however, we will need the bright young and energetic workforce to be able to make the products within our borders. This comes with many challenges. We must promote and attract these workers. We must challenge them so that they become problem solvers and innovators. In order to compete globally, we will need to automate and employ the most advanced manufacturing techniques in the industry. There needs to be high level solutions for the next generation. This leads to the potential answer being that the next generation must be the SOLUTION GENERATION! In the next chapter, we will discuss what this concept means, looks like and how America's Greatest Champion could be this generation!

THE SOLUTION GENERATION'S BEDROCK

"I know I'm right. I know manufacturing is vital to all of our futures and my children's futures when I'm gone. I know that without our manufacturing base, there will not be a U.S. in fifty years. That's not an overstatement, it's a fact."

– Greg Knox, President Knox Machinery and CNC CZAR

Our world has changed. Young people have a much more challenging time than the baby boomer generation did as children. We need to look through our children's eyes and see what they see, taking us out of the larger equation. This next generation is our future. We have so many things that are going to require change, and someone will need to step in and change the culture. It could very well be that the solution is found in the hands of our next generation, hence the title of this chapter.

Additionally, we need to look at the changes that are being presented and how these might help pave the way for manufacturing to come back to new heights. While the new administration is not always making the popular decisions, there is hope that there will be new ground gained on so many other fronts.

WHEELING HIGH SCHOOL STUDENT MEETING
SPEAK UP FOR YOURSELF!

I was given the opportunity to speak to ten brilliant young people at Wheeling High School. They were given the opportunity to ask any questions of me that pertained to engineering or manufacturing. At the time with thirty seven years vested, I thought that would certainly give these high schoolers incentive to ask away. Three of the ten students were very engaged. They were obviously confident, smart and articulate. After a considerable amount of time I said, "Ok you three are done for a while." I then proceed to say that everyone there would be expected to ask at least one question. I told them we were staying there until all students has asked one question. At that point, four others also became engaged. Questions were asked—good questions. I gave honest and well-intentioned answers. This created yet another dilemma: the three students (all three of which at this point happened to be young women) who had still not yet asked questions. I then told them all, "You know that you three could have the best minds, be the brightest engineers, the most creative ideas in the room, but unless you engage, I will never have the opportunity to realize that." I went on to say, "If you are shy, that still is no excuse. You cannot give yourself a free pass. You must present yourself for who you are—not by your silence or inability to speak up—and engage with a future potential employer."

The point of the story is that this generation has many daily opportunities to show this behavior. It is reinforced with each social media tweet, and each and every text or instant message.

As a young person, I needed the social opportunity to interact. It was important to me. I thrived on it. I played all team sports, not individual sports. I needed the opportunity to interact with others. Not all young people in today's world have that need, that thirst, and technology does nothing to promote it. We live in a time that condones being a recluse. Young people would rather text than pick up a phone and make a call. Where are our future public speakers? Where will our future leaders come from? My hope is to speak to young people across the country and show that my generation believes in them, and mentoring is a big element to that.

ILLINOIS HIGH SCHOOL DISTRICT 207
COLLEGE AND CAREER READINESS

I am a big believer that young people are getting away from certain skill sets that set them up for success, not only in school, but also for their careers after school. Being college and career ready is paramount for our young people. Here in the suburban area of Chicago, District 207—which includes Maine West, Maine East and Maine South High Schools—started a new program that they employed for all students in their district. The program was implemented as a Satisfactory or Unsatisfactory score on each student's progress report, but not their transcripts. This was a College and Career Readiness Skills Program[1] that included four key skill sets:

1. RESPECT: In the world that I live in, which is manufacturing, older people really want to help young people succeed. However, when young people come in and don't give respect (or pay attention), and are not curious or sponge-like in terms of learning, that really disappoints the older people in the mentoring process.

2. COLLABORATION: John D. (Maine East Class of 2013) says one of the most underappreciated skills is that of collaboration. When you try to make a name for yourself, what you do not realize is that in order to get there, you need to rely on a lot of other people. Having the ability to communicate your ideas effectively and understand the ideas of others—ultimately work with them—is something that is going to get you far in any setting—career, college and even just life.[1]

3. HABITS FOR SUCCESS: Joe C. (Grant Merchant Services) says that you really need to be ready for what you are doing and be present mentally and on time. You need to be ready to go, which seems simple, but is huge to step up your future in the company.[1]

Employers want people who are accurate. We want workers who are diligent. It is a known fact that if you set goals, you will attain more success than by merely setting those goals.

4. TIME MANAGEMENT: Jane A. says that there are 168 hours in a week. How am I using each slot of time? I need to break those down into chunks of time.[1]

Lastly, I feel that young people need to find something that they are passionate about when they are young. In my case, I hope it is engineering or manufacturing, but if not, that's okay. If they find something early in the game, whether it is their freshman or sophomore year, something that they are excited about, all these four skill sets will pretty much take care of themselves. It won't seem like you are trying because it is something you enjoy and you love.

MORE CONVERSATION WITH NICOLE MARTIN
AUTHOR OF "THE TALENT EMERGENCY®"

Gen-x-ers are happy to go email all day long, and millennials will do it in text and won't pick up the phone to call you and talk about it unless it's significantly apparent that it's important to them, but that doesn't mean that it's not important to you. So, I think communication guidelines have become pretty much common practice or should be common practices in businesses to outline kind of, "Okay, this is what's appropriate for a phone call, this is what's appropriate for a face-to-face meeting, this is what's appropriate in email," or response times in email, for example. Don't email something unless you're expecting not to hear from me for two days, you know?

Well, what's happening is that the younger workforce is being put in positions of greater responsibility sooner than they've had opportunity to adapt and be trained for. That's a result, again, of the talent emergency. So part of it isn't just the soft skills, it's that we're not training people or setting them up for success properly.

TECHNOLOGY AND OUR CULTURE – THE GOOD, BAD AND THE "GOOGLE-Y"
ARE WE READY FOR THE "MANUFACTUR – LENNIALS"?

In his article entitled "Will Millennials Change Manufacturing?," Steve Minter of Industry Week Magazine takes on the subject without pause. He writes: *"The largest generation in the U.S. is taking its place in manufacturing—and the experts are betting this tech-savvy cohort is ready to stir things up."*

He contrasts the outdated perception of manufacturing—"dark, dirty and dangerous"—with the type of industry millennials want to be part of: "attractive and safe, innovative, even cool."[2] This is important to note because, as Minter points out, millennials are the "largest (group) in the United States—83.1 million, according to the U.S. Census Bureau versus 75.4 million baby boomers. Not surprisingly, millennials also make up the largest share of the American workforce—one in three workers is a millennial, the Pew Research Center reports. As baby boomers leave the workforce, and millennials make up a more significant part of it, many manufacturers believe that this generation will change manufacturing."

WILL WE FIND ENOUGH TO KEEP THE SPINDLES TURNING?

In Chapter Two we stated that according to both Deloitte and the Manufacturing Institute over the next ten years that it is likely that nearly two million manufacturing jobs will go unfilled.

Minter writes, *"Many manufacturers, even in successful, growing companies, clearly worry that millennials will not be attracted to manufacturing."*

CAN THE MILLENNIALS GENERATION BE THE SOLUTION?

Earlier in Chapter four Nicole Martin speaks about how every business manager needs to have the ability to respond to differences between generations in what she refers to as Generational agility. Minter goes on to point out,"It would be foolish to think that a group of more than 80 million is monolithic in their characteristics, but many experts say manufacturing can count on millennials to be open to innovation and lead the charge for change."

This why they could be become to be known as the "Solution Generation" for the dilemma that the manufacturing world in this country finds itself grappling with. Only time will tell. We all need to jump in and give it a fighting chance by making sure many of these young people have the choice, because they know of the opportunity. We cannot afford them not knowing of the needs, as a reason for them choosing other careers.

A CONVERSATION WITH BRAD SCHNEIDER
"LIFE IS NOT A RACE"

Over the course of the years, I have been able to make acquaintances with a variety of people, many of whom believe in my passion for manufacturing. Once such person is Illinois Congressman Brad Schneider. We were able to meet at a variety of manufacturing companies in his district in Illinois. Brad invited me to speak in Washington DC to the Small Business Committee of the House of Representatives on the difficulty of finding skilled workers for small businesses. To date, that was one of the more fascinating experiences for me, personally. Here Brad speaks to the vision of the book and how we need to join together for a common good.

There are four key pillars for business success: Number one is a business model. Number two is the people you have, your talent. Number three is the capital used to give your talent the resources they need to succeed. Number four is a stable business environment.

The first one, the business model, that's up to each entrepreneur, business manager, and company strategic planner to have a model that fits a market and exploits an opportunity. But government has a role to play in the following three.

With access to talent and access to capital, there's a partnership between the private and public sector. We need to make sure we're educating our young people, giving them the skills and lessons they need to succeed, the confidence to reach out and push their limits and explore and find new opportunities.

As far as capital, we need to be constantly working to make sure that we have both effective and efficient capital markets, that businesses are able to access the capital they need, whether that's through equity or debt and any combination thereof.

The fourth one is really on the shoulders of policymakers of government: local, state and federal. We need to work and do a much better job of creating a stable business environment to give entrepreneurs and business owners the confidence to invest, and to give workers the confidence to expand their learning and take on new opportunities. A way of looking at that is trying to clear the path. Set the path for American businesses to succeed, and clear away the obstacles that stand in their way.

Specific things we can do include making sure that our tax code provides incentives for inventing things here, R&D, product innovation, making things here, production credits, and shipping them around the country and around the world. There are very specific policy things we can do. We also need to be advocates for manufacturing. Going to schools and talking to young people about the opportunities, as you and I have done, as well as working with our veterans to translate their skills from the military into the private sector. The skills they learn in their military career are directly relatable but not necessarily directly translatable to the private sector..

We're reaching out to the youth. I recently visited twelve companies as part of my "Made In Illinois Tenth District Tour." Every one of them talked about the difficulty they have finding qualified, capable people and the opportunities they see if they could get the people to fill the spots.

Yesterday, I started my day at a pre-k through second grade school which, this year, was converted to a STEM immersion culture. They showed off their epic accomplishments working with Lego robots. It was really extraordinary to see what these first and second graders were able to do using technology (an iPad, a Lego product, and learning to code). I was impressed with the passion these young people had for what they were doing on these fairly straightforward and simple tasks, although they were doing it with a fairly complex set of technology. If we can work with young people and show them the opportunities that lie ahead, the technologies that are being developed today and potential for technologies that we haven't even thought of that will be developed tomorrow, and say, "The future is yours," we'll go a long way in making sure that these kids—who I saw yesterday as first and second graders—continue to pursue not just STEM but STEAM (to add the arts in there). They should also have a broad-based, well-rounded, sophisticated, 21st-century education that makes sure they have the skills that we need them to have to grow our economy.

Life's not a race. As you pursue your education, you're not going to finish your education in four years. Even if you get your diploma in four years, you're going to be educating yourself throughout your career. It may take longer than four years. Some people may take an alternate path. There's not one way to success in the 21st century economy.

What's necessary for success in this economy is that you're constantly learning, you're constantly developing new skills. You're taking advantage of opportunities as they're presented to you. As I talk to young people, there are so many ways, if you're looking to get a degree and you're worried about finances, there are opportunities to go to community college, get an associate's degree within two years at an economical cost before going on and getting the bachelor's degree at a four-year degree school. I'd like to see getting to a point ultimately where someone is able to work maybe ten hours a week in a work-study job and graduate with little or no debt. The debt that they do have is something that they should have full confidence that they can repay within roughly ten years from graduation without excessive burden. When I talk to young people, I tell them, "There's nothing that should be able to stand in your way of getting the education you want."

We should be striving to make sure the United States is a nation where we invent things, make things here and ship them around the world, promoting innovation and creativity, protecting intellectual property and opening new markets. Manufacturing is a great career full of opportunities. It's going to challenge you, but you need to understand math, understand science and understand the arts. It's that well-rounded education that's going to help you succeed in a manufacturing environment. We have to make sure parents understand that manufacturing is a great opportunity. A master toolmaker can now make a very comfortable salary. There is opportunity to apprentice and learn the skills from that master, become a master yourself, and have the opportunity for career security. Every parent should want that for their kids, if that's where their passion takes them.

A CONVERSATION WITH GREG KNOX
THE "CNC CZAR" REACHES OUT TO MAKE
A DIFFERENCE

A number of years ago, while representing a builder, I was asked to serve on a distributor board whose function was to hold the builder to standards that their customer base felt were important. Another principle that was on the same board was a machine tool distributor owner by the name of Greg Knox, whose business is based in Ohio. We both share some of the same values and passion for our industry and our customers.

Greg has become a bit of an internet whiz kid, with some very noteworthy connections. I am fascinated with Greg's stories and how he came to proclaim to be the CNC CZAR!

(CNC CZAR) What can you say about that? That's a moniker I came up when the last administration was reigning and the President of the United States was making a bunch of above-the-law czars. I did a mock piece on my website that showed photoshopped Obama patting me on the back, and I was at a mic and had announced Greg Knox as the CNC Czar. I talked about how making me a czar was really the right thing to do, and shutting down all of the other CNC machines who were competitors in Ohio was good for the customer. Its aim was to point out how silly it was and how far we were getting away from the Constitution.

A LETTER THAT CHANGED A LIFE AND MORE . . .

I started in machine tools back in the '80s in Detroit. I started my own company, a machine tool dealership in 1996 in Cincinnati, because my wife and I—both New York expats, decided we wanted to raise our six children in Ohio because of the family-first culture and morals typical in this part of the country. We sell and distribute CNC machinery. We call ourselves a CNC service company that happens to sell machine tools. We take a very service-first approach at Knox

Machinery. As a dealer, we eat drink and live everything in our territory. We know our customers and after 22 years, they know us pretty well. A number of years back, I came into my office one morning at six o'clock and there was a plea from Detroit to call our local and state politicians and let them know that we backed the automotive bailouts.

My fingers just starting whirring at the keyboard. I looked up, it was almost eight o'clock, and I just said, "Wow. I've got to go sell something." I was replying to their email and letting them know in no uncertain terms that I did not agree with the government stepping in to bail out companies that were essentially bankrupt because they put themselves there. Before I hit the send button, on a whim I copied my mom in new York, who has been a housewife her entire adult life. She said, "Greg, can I share this with my friends?" I said, "Sure." Her friends are essentially a bunch of knitting buddies across the country.

A week later, I got a call from someone at NASA in Florida and they said, "We just wanted to call and congratulate the man who wrote the letter." I said, "That's great. What man and what letter?" They said, "Didn't you write a letter to the president of GM?" I said, "How on Earth do you know about that?" The guy said, "Because I just saw it on Fox News." Within minutes, all six lines of Knox Machinery lit up. They stayed lit up for about six months. We literally had hundreds of thousands of emails, gifts and care packages, letters from every walk of life including politicians, and academia at virtually every major establishment (including Yale and Harvard) radio stations and TV crews. After over two decades in manufacturing, I had become an "overnight sensation."

Next thing I know, I'm standing in front of tens of thousands of people at Tea Party rallies across Ohio, saying to myself "How did I get here?" All I did was just tell GM, in no uncertain terms, both you and this country do not deserve bailouts for corruption, greed and laziness. We need to get back to the principles that made America great. That's where my public career was off to the races.

Perception is reality. I've really had a passion for trying to correct the false image and perception of manufacturing for decades now. For too long, our nation's trade schools have been viewed as something for "the less intelligent kids who can't do anything else" and nothing could be further from the truth—if you think about it, where did virtually every NASA engineer and "rocket scientist" start? That's right—manufacturing, and most likely, a trade school before that. That is not to say that every manufacturing candidate needs to have rocket scientist IQ—like every other field, there is a spectrum of opportunities which require varying levels of knowledge and expertise—but make no mistake—the greatest middle class in the history of the world was created over the past century in this country by manufacturing, mining and agricultural jobs—most filled by your "average Joe."

" *I'M YOUR HUCKLEBERRY!*"

Fast forward to last year, when I got a call from the White House. They reached out to me because Vice President Pence wished to visit the state of Ohio to discuss manufacturing.

"Mr. Knox, we understand you're the go-to guy in manufacturing in Southern Ohio." Interesting... They said, "Vice President Pence wants to make a trip into town, and we were wondering if you would be willing to give us three potential sites for a visit. They have to meet X criteria. There will be a manufacturing round table, and we'd like you to pull together a small group of leaders in manufacturing to discuss your issues with the vice president. Would you be willing?" What I said verbatim was, "I'm your Huckleberry!" (A famous line from the 1993 movie Tombstone with Kurt Russell and Val Kilmer)

THE MEETING – TAKE ONE

Between the time of that call, and a few days later, when the meeting actually took place—the nature of the meeting had changed from manufacturing to health care, but the White House asked me if I would be a part of the roundtable discussion to represent the voice of small business, so of course, once I again I replied, "I'm your Huckleberry."

I do not pretend to be an expert in the healthcare field, but as an employer who has literally paid seven figures for my employees' plans since starting my company over two decades ago, I certainly do have some opinions.

I was (uncharacteristically) quiet for most of the meeting, really saying nothing, until I realized that there was time for only one more speaker, at which point I got the vice president's eye and motioned to let him know I was now ready to speak.

I told the vice president, and those in attendance, that I felt the U.S. healthcare system had been broken and declining for a long time in this country and gave them some thoughts, from a simple man, on some things I thought could help (tort reform, "loser pays," adopting a model more similar to other forms of insurance, etc.)

After completing this portion of my talk, I then looked the vice president right in the eye and said "Mr. Vice President, now you're not going to leave the great state of Ohio without you and I talking a little bit about manufacturing this morning? At which point, the vice president lit up like a Christmas tree. I said to him, "If it wasn't for America's ability to go from making typewriters and bicycles to rocket launchers and tanks in times of war, we'd all be speaking German or Japanese right now. Our national security and our manufacturing base are necessarily linked at the hip, and the erosion of one necessarily equates to the erosion of the other...for too long previous administrations have failed to appreciate this fact—and that is at our peril."

I told the vice president that growing, mining and manufacturing are the bedrock this nation was founded on, and that this nation needed to get back to the values, work ethic and appreciation of these industries that we had when they helped to make this nation the greatest powerhouse and economic force in the history of mankind... I said a lot of things—and I could tell, by the way, the vice president was listening that he really "got it."

THE MEETING TAKE TWO – YES, MR. PRESIDENT...

That night I'm driving home, and I get an email which had the presidential seal on top which starts off by saying, "Dear Mr. Knox," (WOW!!!) "Going back to Washington tonight, all Vice President Pence and Dr. Tom Price could talk about was how passionate and eloquent and knowledgeable Greg Knox was. They wish that he would travel around the country speaking with them."

I figured that was really neat, and it was over. Two days later—when my phone rings—it says restricted number, and when I answer it, I hear, "Hello, Greg, this is the White House. The vice president talked to President Trump, and he would like to fly you to the White House to meet with him in the West Wing Monday morning. We're wondering if you were available?" Without missing a beat, I said, "Well, I will have to check my calendar...YES, I'm available!" (Which created quite a bit of laughter in the room).

Then there I was, a few days later, flying off to Washington. I got to meet with the president and Vice President Pence. The format was a listening session on healthcare once again. Then when that was over, I got to talk to both Vice President Pence and President Trump about the things I said about manufacturing. I got to go into the Oval Office with President Trump.

I'm not going to hide my candle. I know what I know. I know I'm right, I know manufacturing is vital to all of our futures and my children's futures when I'm gone. I know that without our manufacturing base, there will not be a U.S. in fifty years. That's not an overstatement, it's a fact. We who are in the know, can't hide our candles. Manufacturing is still the backbone of this nation, and we all need to change the archaic misperceptions that keep bright young talent from considering jobs in this outstanding, lucrative and rewarding field.

God bless America.

Nicely said indeed. In this chapter, we have discussed the importance of the next generation, and the wakeup call that all of us got when the economy hit the bricks—bailouts, government policies that added more rules and regulations than a business can take. With all the reshoring that the current administration is advocating, WHERE are we going to find the workers for making the products our companies are selling? In the next chapter, we are going to make a case that one of the first places to look is to the female gender to make the quickest impact. With the #METOO revolution, there is a newly inspired sector ready to take up and take a stand for unclaimed opportunities.

WOMEN IN MANUFACTURING – CHANGING THE LENS OF WHAT MANUFACTURING IS

"I think when I started my front-end journey around women in the industry, I went to look for stories . . . I went to find the stories of the machinists and the welders and the quality technicians, and they didn't exist."

– Jennifer McNelly, Former President of the Manufacturing Institute

There is a revolution happening for women in so many ways. Women deserve to have the best opportunities that we have to offer. Our mothers, daughters and grandmothers have always given us their best. It doesn't take long to find signals of what is coming in our culture. Turn on the radio and listen carefully to the words of Rachel Platten's song "Broken Glass," and you will know and believe, that changes are coming. Changes for the good, for all.

There is an opportunity for women in this country. Our culture is ready to embrace it. Too many men have taken advantage of positions of power and the #METOO movements will empower women to rise up. All of us will benefit, even manufacturing. Forty seven% of our working population is of the female gender. Yet, we are ignoring a huge resource in employment, specifically manufacturing employment. We need to empower women—especially young women—to pursue STEM-related fields of study and ultimately careers. Men need to be more encouraging and accepting of that truth so they help propel, not hinder, women from entering male-dominated fields.

Why are there so few women in manufacturing? It is a rarity that I talk to a woman involved in the manufacturing or engineering part of the business. I married the smartest (and prettiest) woman I ever dated. I just could not think of spending any significant amount of time with a woman who did not challenge me intellectually. I never claimed to be a rocket scientist, however, I did engross myself in a lot of engineering coursework, AP Calculus, AP Physics and the like. I really enjoyed seeing parts being made, math equations that gave solid answers, and the science of physics and how things work and why. Many men are linear thinkers. Not all the answers come from a linear process. Many women think much differently. There is a tremendous amount of good that could come from looking at things differently—not to mention that we have a void of talent in the labor pool. Why not draw from areas of the workforce where we have not done so in the past?

Our family has been blessed with strong women—from my mom, my wife, both my younger sisters, my cousins, and my daughter, (and daughter in-laws). Having said that, I don't understand why there aren't more women in our industry.

Are men too biased to allow them to become significant in our manufacturing companies? I would like to think that is not the case. I do know that when you meet a woman in manufacturing, she is most commonly strong and very determined and driven to succeed. There aren't enough stories to tell, however, I will do my best to share those who have inspired me!

PARENTS – ENCOURAGE YOUR DAUGHTERS

This past fall, I went to visit a machine shop closing that was a GE division where they machined turbine blades. We had sold them a very high end CNC surface grinder that I was interested in.

When I arrived, a young woman answered at the window. "I didn't know you were going to be here! At first I thought someone was behind me, until I realized the lobby was too small for someone to have walked in behind me. "Oh, OK," I said, in a confused manner. We exchanged small talk until she escorted me to the shop. She asked someone else to take me to the machine. On the way back, I asked my chaperon, "Help me out; what is her name?! I know that I know her, but I cannot place her name." Carrie, he said. "How about her married or maiden name?" I inquired. Kurczynski, I think he said. That must be her married name, but I still could not figure it out.

As I was taking my photos and videos of the machine, she walked back to check on me. "Help me out here," I said. "I know that we know one another, but how? What is your maiden name?" "Johnson," she said. "Ken and Muff are your parents!" I loudly said. At the same time, we said "Kingswood!" That is the church that Kathy and I attend and have known Carrie and her parents for Carrie's entire life. I knew Carrie went into engineering, but I rarely had a chance to talk to her about her interests, her education or career. Shortly after our impromptu meeting, I called and asked her to be interviewed for my book. I asked her how she made her way into engineering, then ultimately manufacturing.

A CONVERSATION WITH CARRIE J. KURCZYNSKI
AT A YOUNG AGE I WAS TAUGHT...

"What you did and how you did it had nothing to do with how you peed."

It's hugely related to my parents. When I was twelve, we rebuilt my dad's motorcycle. It wasn't our first project, he was always tinkering with something, but it was the deciding project when my dad kindly insisted I should consider engineering—telling me that I would love it. When I reached high school, my mom helped point out all the tech classes so I could better understand the school side of engineering. Since math and sciences had always come easy to me, she didn't want me to think it would all be easy.

When I finally got my driver's license, my dad wouldn't let me take the car out until I could prove I could change the oil and change a tire. (Ironically, the only time I have ever needed to change a tire was on my way to a wedding as the maid of honor. I was already dressed for the ceremony when I heard the tire blow. Lucky for me, it was right in front of a mechanic shop, and they offered to change it before I could assess what was going on. I knew I needed to let them as my heels and dress were not going to survive me doing it myself, but I called my dad and vented my frustrations at needing to accept help when I could have handled it.)

While I liked all the hands on projects, enjoyed my technical classes and found that math and science came easy, I was diagnosed with ADD and a slightly form of dyslexia, mostly with numbers, at young age. In high school, I was able to control or work around it with minimal accommodations. Sometimes I would lose points on tests or homework because I switched a number here or there but nothing that affected my overall semester grades. But, I was worried about getting in over my head in college. With the help of my parents, we were able to look into what accommodations were available at a college level, and it turned out that Purdue had many more resources available than I even ended up needing, but they were always there if I asked.

Our (high school) graduating class was 1065 students. If I had to guess, maybe, easily 10% of the women had the ability to do it (go into engineering). I can only remember three or four who did—there could have been more. After college, a lot of companies tried to get me to go in the direction of technical sales. It would have been twice the salary, but I couldn't imagine not getting dirty on the shop floor or sitting in some office eight to ten hours a day. It almost felt like they thought I enjoyed getting dressed up in the suit and heels and putting on makeup. No thank you! I had spent my last year of college career on campus twelve to thirteen hours a day, bouncing between classes, labs and the machine shop. Sitting still just was not for me and heels weren't for me (well, not boring back ones and certainly not every day). When I started learning what a field in engineering was, I was hooked. There was no other option for me and not once in any of the conversations did anyone care that I was a girl. All they cared about was that the long hours, hard work and technical issues wouldn't stop me from pushing forward.

I would tell young women to drop the notion that female has anything to do with what makes you happy or what you want to do in life. Don't compare yourself to the guy down the street or another girl or your siblings. Do the best you can do, and do what makes you happy. Ask questions! Just because no one else has asked something doesn't mean they aren't wondering also. You would be surprised how many people just don't want to talk. I don't know if I never cared what other people thought or if at some point, I just stopped worrying about it, but I've always sort of ended up doing my own thing (for better or worse). You can come with me or not.

I was lucky. I was brought up to believe that what you did and how you did it had nothing to do with how you peed. Maybe that is a little blunt, but my work and work ethic was what counted. I wasn't always the best, but I worked my butt off. In school, it was the same way. I didn't always have the best grades, but I worked hard. And now I work for a great company and they support me not only as an engineer but also a mom and a wife—no need to pick one over another.

CARRIE MEET MOLLY! A GENERAL ELECTRIC COMMERCIAL

Right around the same time of our meeting, I saw the GE commercial about a young woman inventor named Molly. As I watched it, I could not help but make the correlation between "Molly," in the GE commercial, with the nonfictional real life character of "Carrie," working for GE. I texted Carrie and told her about the commercial and the uncanny connection of her working for GE making turbine blades. Wow, some coincidences are just too timely to be anything short of—"a sign." GE has shown up as a leader in trying to make a statement in commercials for manufacturing and women in its workforce.

A CONVERSATION WITH DANA WARD

This empowerment of a young daughter is what is needed for so many young girls. Hopefully, there will be more women entering into the ranks of STEM or STEAM related fields going forward. DANA WARD (whom we talked about in Chapter Nine) tells about her parents' influence in her path to success.

I think it is key to understand that failure is just a normal part of life, because as an entrepreneur, you're going to experience failures of varying degrees throughout. You need to be okay with it or work towards accepting it. I try to quickly address an issue, learn from it, and move on. You cannot allow failure, or even fear of failure, to paralyze you. No one has time for that!

Something that you should always try to make time for—and this is just me preaching, but for others who are motivated by value, it can be helpful too—is to support others in their endeavors. At some point, you'll need help from your network, friends, family and even strangers, so put out what you'd like to receive. If you support someone for who they are, ask how you can support... or if you like an idea or product, then buy it.

I give a lot of credit to my parents for helping structure a healthy mindset;

they really were responsible for teaching me at a young age that I could do anything that I wanted to do. I mean, I'm pretty sure that if I decided that I wanted to become an astronaut and go to space tomorrow, I would find a way to make it happen (even if logically and logistically that means buying a ticket for when a public flight is available). Find that confidence and hold on to it tightly. Inevitably, you'll need it.

STEPPING UP TO ENCOURAGE GIRLS AND MINORITIES INTO STEM

Colorado certainly seems to keep surfacing as a state that is doing all the right things when it comes to STEM and getting more young people involved in coursework and activities. Monte Whaley writes about various contributions that are making a difference in their state.[1]

In Denver public schools, $10 million in bond funds are supplying laptops for 9,000 students this school year. The effort is aimed at encouraging girls and minorities to pursue careers in STEM (Science, Technology, Engineering and Math) fields, DPS officials said. Many businesses, nonprofits and individuals are stepping up to help cash-strapped families meet their back-to-school needs. More than 400 volunteers donated their time to help run the Action Center school supply giveaway in Jefferson County. They sorted supplies and stuffed backpacks and then formed a well-oiled machine to check IDs and match students with the appropriate school gear during the week-long event. One of the biggest donors to the Jeffco effort is local home builder Cardel Homes, which began donating as many as 3,000 backpacks a year to the Action Center in 2014.[1]

WHY OUR YOUNG GIRLS RARELY GO INTO STEM RELATED FIELDS

What is the role of the parent in situations where female students and students of color aren't getting the support they need at school to pursue science? Dr. Mae C. Jemison, the first African American woman to go to space, is interviewed at the Sheraton Hotel.[2] (The Seattle Times)

Parents can play a role. But schools need to get their act together. I'm sorry, but I'm not going to give schools a pass. My mother was a schoolteacher, she would challenge me. My dad hunted and fished, he did construction and contract work. I hung out with my dad and his friends when I was six or seven. I played cards with them. I learned how to count cards. They thought it was great—that this little girl could count cards. So I was very comfortable in positions where I was being challenged, even around guys.... It's not about parents pushing kids. It's about allowing them to explore their interests and letting them know that science is important. Another thing parents can do—uh oh, this is a tough one: stop requiring that your daughters be nice and clean and pretty with their hair done. You can't do good mud pies and have your hair done and stay clean.[2]

To Mae Jemison's point, I coached a lot of soccer, both young men and women. When the girls were younger, they were afraid of getting dirty. They, of course, had been told by everyone—parents, grandparents etc.—to always look your best, which of course usually does not include getting messy or dirty. Well in soccer, that takes on a different definition. When the weather conditions were dry and hot, all was good. Put in a little rain and mud, and there is where the problem started. Finally, after a few seasons, I got wise. Anytime we arrived at a field that had standing water or was wet in some manner, I would call the girls together and tell them prior to the start of the game to run out to mid field and dive in the grass. "Everyone roll around in the standing water and mud. You are going to get dirty anyway, so let's get this out of the way." Not only did it eliminate the girls' concern about getting dirty during the game, it sure intimidated the other team who then knew we meant business!

A CONVERSATION WITH JENNIFER MCNELLY

THE MANUFACTURING INSTITUTE and STEP AWARDS FOR WOMEN

When I joined FLATE I got to know some pretty interesting people. One of my NVC (National Visiting Committee) counterparts was Jennifer McNelly, who was at the time the president of The Manufacturing Institute, a great source of statistics. She founded an award called STEP of 100 women in manufacturing. She's a believer of women being engaged in leadership positions in manufacturing. Jennifer has spent most of her career advocating for technical careers.

I have spent the majority of my career in manufacturing, and during my time at the Institute, we published a lot of the statistics on manufacturing and the economy. Finding and retaining a skills workforce remains a top issue for manufacturers. According to the reports published by Deloitte and the Institute, 80% of manufacturers can't find a quality workforce, even if they are willing to pay.

What is the cost of lack of action—lack of change? It has real business implications. Unfilled jobs can be as high as 11% which equates to lost earnings for our nation manufacturers in cycle time, downtime and overtime, according to a report published by the Institute and Accenture.

When you start to look at the challenges and opportunities, traditionally manufacturers think about the next generations. How do they engage in the pipeline of young people, the perception of manufacturing careers, and whether parents are encouraging kids into manufacturing careers? One of the greatest opportunities is to attract new demographics into manufacturing, including women.

Women are 51% of the population, 47% of the U.S. work force and 29% of the manufacturing workforce. I've seen improvements in that number just in my time in the leadership around women in the industry. Women are an untapped labor pool. The question is how do we engage more women to consider manufacturing as a career?

More diverse companies are in fact more profitable. Diversity in the workforce has bottom line impact (that) tends to change behavior more than anything else. You can tie the economics of why change is important.

During my time at the Institute, we went to look for stories of women in manufacturing to share with young girls. We wanted leaders to tell to the next generation about the opportunities to help build the pipeline for young girls and boys. We went to find the stories of the machinists and the welders and the quality technicians, and they didn't exist.

To help change this perception, at the Manufacturing Institute, we founded the "STEP Ahead Initiative" which recognizes women in science, technology, engineering and production. STEP was unique in its approach because it told the story of women in leadership, not just the C suite—but the women at every level of the manufacturing workforce.

Women lead at every level every single day. When we share their story and empower them to act, we amplify the reach to inspire others into manufacturing careers. In 2017, on the 5th anniversary of STEP, Deloitte and the Institute published a report on impact and what they found was the 500 women who had been recognized, reached and impacted over 300,000 women and girls. That's the personal responsibility and accountability of paying it forward.

With our work on STEP, we recognized the importance of executive engagement. Women do not succeed without the support of men in the industry. Terry, you are a great example of a true Champion. Having men be advocates, sponsors and mentors challenging us to be more than we think of ourselves, is really important in this equation too.

I think about the trailblazers in manufacturing. Rosie the Riveter, Mary Barra, and countless others who are the shoulders of greatness the rest of us stand on. For manufacturing to remain competitive, we need a quality workforce. To attract that workforce, we need everyone engaged.

Everyone in manufacturing is part of the solution to our skills gap. To the women and men that read this, be proud, stand tall and talk about what you do every single day.

I am proud to be part of the manufacturing economy. Manufacturing builds communities and strengthens the economy. Be proud of what you do, what you make, and share your story with anybody that will listen. Even those who don't want to.

A CONVERSATION WITH ALLISON GREALIS
WOMAN IN MANUFACTURING ASSOCIATION LEADERSHIP

Allison Grealis is founder and president of Women in Manufacturing (WiM), a national trade association focused on supporting, promoting and inspiring women in the manufacturing sector. She is also the vice president of membership and association services of the Precision Metalforming Association (PMA), a full-service trade association representing the metalforming industry. I called Allison to get her perspective about her views and the organization's vision going forward:

Women in Manufacturing is the only national trade association serving the interests of women in manufacturing year-round. Our mission is to support, promote and inspire women, specifically with a focus on education, training, networking and providing resources to women who are in the industry.

We're focused on women who have already chosen careers in manufacturing and work to help them stay in their profession and have opportunities for advancement. Currently, there are 1.5 million women in manufacturing, which is a very large population, but sadly too few of them are in the C-suite or in senior leadership roles. One of the key things our organization is challenged with is how to serve all women in manufacturing including those on the shop floor who may someday aspire to rise in their organizations. As

227

these women in production often have less opportunity to leave the shop floor for training and networking, we are working to devise creative ways of using technology to reach this important population.

We're focused on women who have already chosen manufacturing careers and helping them advance, stay in the profession and have opportunities for great careers and advancements in their organizations. We look at the current data, there are 1.5 million women in manufacturing. That's a very large population. One of the key things that we, as an organization, are challenged with is how to serve all women in manufacturing, and that's our aim, to not only serve women in the C suite and above, but how do we get to helping those women who are on the shop floor who aspire to someday be in management or to be running a plant? How do we help them?

One of the biggest obstacles that we still face is the outdated perception that much of the general public has about manufacturing. We work to promote the reality of today's modern manufacturing and the amazing careers that this industry offers.

I think that education about opportunities in manufacturing is critical. That is why our organization is very focused on telling the story of modern women in manufacturing through our "Hear Her Story" series. To date, we have profiled more than fifty women. In this series, we highlight what their lives look like, the skill sets they apply to their work and what led them to a manufacturing career. This storytelling is so powerful, and we believe it can help inspire young women and men to want to pursue a modern manufacturing career.

General Electric is one manufacturing company that has done an exceptional job promoting manufacturing careers as they embark on a journey to add 20,000 women to their workforce by 2020. They have powerfully celebrated women in their company and in the industry. One of their recent campaigns has centered around sharing the story of Millie Dresselhaus, an accomplished engineer and scientist, and how amazing would it be if she was a household name? What if she was as widely known as the Kardashians—how different would our culture be?

Mildred Dresselhaus University of Chicago PhD'58 *Presidential Medal of Freedom, 20143*

Mildred Dresselhaus is one of the most prominent physicists, materials scientists, and electrical engineers of her generation. She is best known for deepening our understanding of condensed matter systems and the atomic properties of carbon, which has contributed to major advances in electronics and materials research.3

One of our long-term goals is to change the perception that moms, in particular, have about manufacturing. We know that moms seem to be one of the chief influencers when their children are making decisions about careers. We are working to more actively promote manufacturing careers with this population as well as to highlight the key benefits and advantages that careers in manufacturing offer. Kids who pursue a manufacturing career are able to avoid the debt that often accrues during pursuit of a college degree, and they are able to earn a high-paying salary and have the opportunity to make things in a sector that strongly powers our economy.

For the past five years, I have to say the Manufacturing Institute "STEP" awards are a big step (no pun intended) in the right direction! Let's emphasize and promote those women who have managed to not only persevere but also succeeded and make them role models for the young women in the future. I have no doubt when Jennifer McNelly was at the helm of this organization, this was one of her main projects starting back in 2013. Kudos to her for leaving a legacy that can inspire young women for decades going forward! After reviewing the list of close to almost 800 women that have been chosen over the last five years, I get a sense of how accomplished these role models are. These award winners represent some of the best manufacturing companies, big and small, in the USA: Caterpillar, Boeing, Harley Davidson just to name a few.

NILES NORTH AND WEST HIGH SCHOOLS

Recently, I was speaking at two high schools in Niles, IL. During the two advanced math classes, I encouraged, as I always do, for the students to ask questions. Here is a seasoned manufacturing veteran willing to answer any question that they could throw at me. You would be amazed at how few questions get asked. However, these two classes opened my eyes to something that I did not expect. I had approximately ten questions asked of me during these two presentations. That in itself was an accomplishment (as I warn them that I will call people out if they do not ask any questions). The biggest surprise was that of the ten questions, seven were asked by young women in the classes! Could it be that possibly there is a change that is coming about? A change that could lead the way to finding the new leaders in the manufacturing world? This would be awesome, and I cannot be prouder of the Niles North and West female students and their boldness and confidence to engage in my presentations! Nice job ladies.

ON THE BEST DAY – WOMEN ONLY CONSTITUTE 29% OF MFG WORKFORCE

With all of this being said, our industry has so few woman as machinists, managers, and owners. Those who have stepped into leadership roles are recognized and rightfully so. The Manufacturing Institute for the last five years recognized soon to be almost 800 women who have risen to the top levels in the manufacturing world. These awards are named STEP awards. Some of the women that I know in this portion of the country have won this recognition. So how are we going to get more women into this exclusive group? In my estimation a mere 15% (or thereabouts) of people in manufacturing are female.

NEW TRIER HIGH SCHOOL

Earlier, I made mention NSERVE, the organization that works with nine local high schools. Martha, who runs the organization, called me about a young woman intern by the name of Alexandra Klieger. She called and asked if I could possibly accommodate an internship for Alex. I said sure, and had her set up the initial interview and discussions. Alex started and did some research for me, while I also would take her on field trips to some local manufacturing companies. Alex was a very bright young lady who went on to the University of Illinois and received a bachelor's degree in Mechanical Science and Engineering. She then went onto the University of Maryland for a masters in Fire Protection Engineering and now works as a research engineer at Underwriters Laboratories. (UL)

ELK GROVE HIGH SCHOOL

I serve on a number of advisory boards at both the high school and community college levels. One such high school is Elk Grove High School. During one of the meetings, I heard from the STEM/Project Lead that, in fact, they had great success with an introduction to engineering class with only female students. The reason that they decided to have a female-only class is that their male counterparts would often kid them and many would lose interest or drop out of the class. Put them together where they are encouraged rather than made fun of, and they found that they prospered as a group. I offered to mentor this group and present to this class. Additionally, I said I would find women speakers from our industry if that would help. I was so impressed with these young people, young women. They asked great questions—all of them.

Going forward, as I make presentations to other high schools with boys and girls enrolled I have started to see a trend. More young women seem to have more interest, actually ask questions—all the while the young men sit saying nothing. While this is a bit disturbing for the young men in the class, what a testimony that maybe the pendulum has swung the other way. Maybe, just maybe, there is a new breed of young women who will be racing into the engineering and manufacturing positions! These are not only good paying jobs, but these young ladies will earn a premium from an industry waiting and wanting—those who will not be from the "old boys club."

WHEELING HIGH SCHOOL

At Wheeling High School each year I am asked to participate in their Careers and Coffee presentations to students. Since I am sent forty or so students who are interested in the engineering field, I make a habit of going around and asking WHICH field of engineering they are interested in. The first year, 75% of the students would say civil engineering, some electrical engineering. Needless to say, I was disappointed. My field needs so many mechanical engineers, manufacturing engineers and more. Then I got to the back of the room. The class was all but over. There was a young lady who softly stated aerospace engineer. After a short period of time speaking with her, I could tell this young lady was nothing less than brilliant. I took AP Physics and AP Calculus in high school, and no doubt this student was going down that path. She was only a sophomore at that point. I was impressed. She is now attending Stanford studying electrical engineering. No surprise here. This is the type of student manufacturing and engineering is waiting/looking for.

MILWAUKEE AREA TECHNICAL COLLEGE

When I was working with Milwaukee Area Technical College, I would present for their "Heavy Metal Tour." Each year, they have manufacturing companies come in to talk to high school students. By my suggestion, they decided for me to present to the students prior to talking to the manufacturers. By having an understanding of the opportunities FIRST, they will be way more engaged and likely to ask intelligent questions. A young woman approached me after my presentation asking questions. This time it was presenting more of a statement and dilemma. "Mr. Iverson, I want to be a welder," she stated. "That's Awesome!" I said. "But my parents are not sure that they want me to pursue this. What should I do?" she continued. "Well do you enjoy it?" "Absolutely!" she snapped convincingly. "Well then, I would educate them on all the opportunities that are in that field. I know of a woman down at Harper College who is loving her life as a welder and would be a great resource for you." When I returned I made sure to get both of them in contact with one another.

PALATINE HIGH SCHOOL

While working with Palatine High School, I wanted to figure out a way to encourage young people to enter into Project Lead The Way classes. I met with Mark Hibner, their instructor, and let him pick the top ten students to be ambassadors around the school. I came up with a poster program that would be the marketing campaign for the class. The poster would feature photos of young people in their high school activities, as well as photos of them in potential future manufacturing related activities. The tag line on the posters was "They were Champions THEN and they are CHAMPION Now!®" I also designed CHAMPION Now!® shirts with fabric pens provided. Each of the students were given the task to get all of their peers to sign the shirt. The front of the shirt said CHAMPION Now!® while the back coined my Manufacturing Create$™ slogan. Wouldn't you know that two of the young ladies took the challenge and "ran" with it! They got so many signatures that it was obvious they REALLY got engaged in the task. This is the type of program that could revolutionize Project Lead The Way programs throughout the country.

How can we, as a culture, ignore the talents and energy that the female gender offer? When you look at all the percentages, the highest % of the working population in manufacturing that I have found is 29%. We are still missing out on 10-20% of highly talented workers. We need to empower women to have the courage and confidence to enter into STEM related fields.

Fast forward to today when we pick up a newspaper, see a news report on TV, or read about a scandal online. Numerous men in powerful positions are being accused of sexual harassment of fellow female workers. While I cannot judge nor contest the allegations, I feel that a new day has come where women do not have to be treated as second class citizens in the workplace. The time and place for these antics never was, and hopefully will be no more. Young men need to know that this should never be tolerated, and women should be respected.

Going down the line back to the schools—the day is coming when boys are not kidding their girl classmates about being smart, or taking coursework that could lead to manufacturing or engineering careers. Once in their career path, women stick with their chosen field and become managers, and owners. We all benefit. A new dawn has then begun where a double digit percentage of women workers enter into our workforce! The lack of skilled workers has begun to notably shrink, and a big positive change has occurred for employers throughout the USA.

TMA – "MOOVING" WOMEN'S CAREERS FORWARD

Egon Jaeggin, also on the Education Foundation, called me about a speaker for an upcoming Education Foundation fundraiser. Egon explained that he met a new force in the industry—a woman by the name of Karin Lindner. She was trying to make manufacturing sexy! He asked my opinion and would she be a good match as a speaker for the event? Oddly enough, I had already come across Karin, and we had become acquaintances through the writing of her book (*How Can We Make Manufacturing Sexy? A Mindset of Passion and Purpose from the Production Floor to the Executive Suite*) and some brief phone conversations when she interviewed me for her book. I told Egon that yes, she would be great; however, I think that the group should vote together to agree on this. Sure enough, we had a conference call, and all agreed Karin would be a great keynote speaker.

Egon Jaeggin, Gregg Panek and Bob Weisheit were three of the half a dozen manufacturers who I became friends with that inspired me to work closely with young people in manufacturing, education and careers. These men have served as leaders in the manufacturing sector and gave me insight on, not only how to give back, but that we must ALL pay it forward as the saying goes.

Panek Precision (Gregg's company) was talking to me about my cause and the TMA Education Foundation's new direction on being a recruitment force for its members and industry as a whole. They told me that he, in fact, had purchased a table for the fundraiser but unfortunately would not be able to use it. He asked that I fill it, which I told him I would certainly try. Gregg Panek is the father and his son, Brian, came to work for his dad in another successful family manufacturing business.

I thought, what a great time to make use of an open table after meeting with the Moovers. Maybe they would be interested in attending. I thought that we would have to merely have the mentor adults explain the use of the word sexy from our Keynote speaker and the way that she used this term. I called up their leader and after discussion amongst the girls and parents, they agreed to attend the event.

Karin Lindner and I talked on the phone about how she would have the most unlikely of admirers at the presentation. Karin was delighted to hear that a young group of girls potentially interested in manufacturing/engineering careers would be attending and that maybe she could find a way to use them in her presentation.

Flash forward to the night of the event, March 2012. I had brought my father and his wife, myself and my wife, along with my sales and service manager and their wives. The Moovers arrived all dressed to the nines, looking much older than I remembered just a few short months before. I introduced the group to our keynote speaker Karin Lindner. Karin was very impressed with the young women. Their leader commented to me that they were probably going to leave early to get the girls home at a decent hour. I told her that was certainly no problem, we were just glad they came and hoped they enjoyed themselves.

Soon after, we all went into the ballroom and sat down for the event to begin. Karin started her presentation. It was not long into the speech that the Moovers were mentioned. Photos of them appeared in the presentation, a touch that Karin had coordinated. Karin called them on the stage and started to interview each of them. The questions were not rehearsed, and the girls answered them quickly and succinctly.

Later in the evening, I looked to see that the Moovers had probably made their exit, so I thought. What I saw was, Karin signing her book for each and every one of them. Egon Jaeggin waiting for them to take a photo with him. Others in the TMA also wanted their photo with them. Overall the event and the introduction of the Moovers to the TMA membership was a huge success!

A CONVERSATION WITH BUZ HOFFMAN
(#METOO) " IT IS A REVOLUTION NOT AN EVOLUTION!"

Buz Hoffman spoke in Chapter Six about parenting. Here he adds comments about the current culture and opportunities women have in the workforce.

I think the circumstances that are now coming about as a result of this thing that's happening will allow women to be and to do it now. You're going to have the good old boys backing away a little bit. But, they're not going to wield. This whole thing is a power thing. It's all about power. Now that this has happened, if the women are anticipatory, if they're intuitive, this is their moment to rise. They've got to do it. They've got to want to rise. I'll still go back to getting this recognition of all kids, not just women, but men too, of what's out there, what's available. Here's what it will take to do it. Here's what I can expect when I get there. They still have to be made aware of the real economic life is past twenty-one, twenty-two, twenty-three years old. Then, it's up to them to grab it. This is a sexual revolution. This isn't an evolution. What's happened in the last couple months, they have torn down the G-D walls. This is a revolution. The revolution is here. I think you need a whole chapter on this in your book. I'm not sure if this is a Rosie the Riveter kind of thing.

According to Bureau of Labor Statistics data, women account for 47% of total employment, but just 27% of manufacturing employment.

HOW TO CHANGE OUR CULTURE?

There are so many perceptions that need to change. In order for manufacturing, engineering and construction to thrive, we need more people within their ranks. Perceptions need to change. In the next chapter, we look at global success stories and cultures that are different than our own. Maybe by understanding these concepts from other parts of the world, such as Switzerland and Germany, we can then impact our culture and perception in the U.S.

THE CULTURE OF MANUFACTURING FROM A GLOBAL PERSPECTIVE

"Germany has the system, but they do not have the people. Whereas, the United States has the people, but does not have the system."

–Mark Tomkins, President/CEO

German American Chamber of Commerce of the Midwest

There are other parts of the world whose cost of living exceeds ours. Yet, they have found a way to embrace manufacturing into their culture. The U.S. can learn a great deal and can only change by changing its inherent culture/educational platform. To put Mark Tomkin's comment at the beginning of this chapter in perspective, the U.S. Census Bureau—World Population Clock 1 lists the United States population at 328 million people compared to Germany's 80.5 million people. 1 (This is approximately the equivalent of combined population of the states of California, Texas, and Florida.) While projections to year 2050 show Germany losing 9 million people to 71.5, the U.S. is set to gain another 1.4 million in the same time period. Another interesting fact is that India, China and U.S. total populations accounts for 40% of the world population. [1]

A CONVERSATION WITH HARRY MOSER
WHERE ARE THE WORKERS TO MAKE THE PRODUCTS?

Back in the late 1990s time frame, there was a machine tool executive who was reaching out to students telling them how profitable and rewarding manufacturing could be. His name—Harry Moser. I started making presentations at the high school our children were attending so that there would be a heightened awareness of careers that were all but unknown to them.

We need to do a better job of balancing the attractiveness of apprenticeships and technical associate degrees relative to four years on campus. Stop building palatial dormitories and rec centers. Cut university loans for liberal arts. Make technical associates degrees almost free. Subsidize apprentice programs. Place priority on providing great careers and vitally needed skills, rather than four years of social life.

From 2010 thru the end of 2016, we have brought back over 300,000 manufacturing jobs that had been offshored. The number one priority to accelerate reshoring: more, smarter, better trained skilled workforce. A key to recruiting that workforce: visible success with reshoring!

Swiss and German manufacturing wages are typically equal to or higher than in the U.S. Nevertheless these countries have much stronger manufacturing sectors and trade surpluses. Reason: apprentice programs. Large numbers of smart youth trained to perfection. The U.S. has been eating its seed corn for at least thirty years. It's time to plant and nurture the crop of manufacturing technologists needed to prevent the U.S. from drifting into permanent economic decay and irrelevance.

I remember a 1997 Production Machining Magazine article showing a child looking at a CNC machine tool in Europe (EMO Show). At that time, our biggest machine tool show did not allow anyone under eighteen to attend. Over my career, I have worked every even year since 1982. That's a lot of shows!

CHRIS KOEPFER Editor-in-Chief, Production Machining Columns Post: 10/1/1997 [2]

"That's the point of the accompanying photograph. It was taken at the recent EMO show in Hannover, Germany. What I see in this picture is a possible beginning of this young man's interest in metalworking manufacturing. Maybe we here in the States should consider making our machine tool shows available to our children. Perhaps set aside an evening or afternoon for families to visit a trade show. We have a shortage of people with the skill sets necessary to work in our industry. Maybe if we could show our children at an earlier age what we do, like this father is doing with his son, we could instill interest in manufacturing at a time when impressions are easily made. [2]

A CONVERSATION WITH FLUID POWER GIANT, AL CARLSON

Over the years, I made calls on a number of companies in all sorts of industries. Precision machining hydraulics is a key area of manufacturing for machine tool distributors and builders. I was fortunate to sell to one of the industry leaders—HydraForce. Jim Brizzolara is their co-founder, President/CEO. Jim and I became friends over the course of many years. When I started on the National Visiting Committee for FLATE in 2010, I was also fortunate to get to know another hydraulics industry visionary Allen Carlson. Al is the President/CEO of Sun Hydraulics. What an opportunity it was to share a phone conversation with BOTH giants in the hydraulic field. The result was a great phone conversation about the workforce, describing their operations and the future.

Let me just add one thing that might be helpful for the readers. I think of hydraulic cartridge valves as a fluid processor. In the electronics industry, it would be a microprocessor. You could think of the cartridge valves themselves as equivalent to a microprocessor in an electronic circuit. The microprocessor and a cartridge is useless on its own. If it's a microprocessor, you have to put it on a printed circuit board. The printed circuit board that's in the hydraulics world that we operate in is called the manifold. It provides the interconnection for the other components in the circuit, much like a printed circuit board would do if it was an electronics product.

Besides Sarasota, Florida, we manufacture in the UK, Germany, Korea, and China. The German culture puts more emphasis on the job training, apprenticeship, and as a career path for people of Germany. For a relatively small country, you end up with world class manufacturing, especially in automotive (BMW, Audi, Volkswagen). They're very strong in other industries as well. It comes from their technical education that starts at a very young age. They're encouraged to go forward with apprenticeship programs and technical education programs. I would put Korea not far behind Germany in their emphasis on manufacturing and technical education. Korea is a very very strong country that has come from the Korean War in the 1950s to where it is today. It's an international power when it comes to manufacturing and engineering technology. I would probably put the U.S. in third place, in that category. Then, the other countries that I've seen are distanced from that.

A lot of the tool and die makers have moved. The big reason for the movement early on was cost. U.S. companies were outsourcing their dies to places like Taiwan. This happened fifteen to twenty years ago. They've built up a base of capability. It doesn't mean that we don't have capability. It's a global environment out there, so if some part of the world can do something as good or better, it doesn't mean that there's a problem with our system. It's a global environment that we're playing in.

My view is that if manufacturers have a skilled workforce problem, I would ask, "What are they doing about it?" If you say that you can't find skilled workers, are you working with the schools, the colleges, universities? Do you have employees spending time in the classroom with young students? I don't buy the philosophy that we have a skilled worker shortage. Generally speaking, the skills that employees need, they can learn once they start working in a particular area of a company. They will learn the skills. What I find missing is the work ethic, the communication, and the collaboration. I call those the underpinnings of a successful employee: the attributes of the person you want to hire. Skills are secondary. Companies need to take ownership of that.

I meet with just about every new employee. The first question I ask them, "Are you looking for a job, or are you looking for a career? By the way, if you're looking for a job, you ought to go down the street. If you're looking for a career, this is the company you want to be with." It speaks to the point of providing new hires career opportunities, not just jobs. One of the concerns I have with the millennial group is that they take that to the new extreme. The millennials are not looking for a job. They're not looking for a career. They're looking for an experience. They will bounce around from company to company, looking for the next new experience. That's really problematic with hiring from the millennial category.

I think that there's maybe some proof to the statistics. I think it speaks to the fact that we don't do a good enough job of collaboration on the front end of product development. The print doesn't go to the shop floor until all of those little things are ironed out. Product development also includes manufacturing development and marketing development. As a country, we get too stuck in our silos. You throw things across the wall.

Back in the 1960s and the 1970s, the U.S. automotive companies felt that they were good enough. They stopped getting better. Along comes the Japanese, and the Germans, and blew right by them as they were quite complacent with their, "It's good enough."

Just one closing thought: for any society, any country, any state, any region, there are three core industries that provide essentially all the jobs, plus the service industries that support them. If you think about it, it's farming, mining, and manufacturing. Everything else lives off of one of those three. All three of those core industries. If you want to develop a society that's affluent, manufacturing plays a very significant role.

A CONVERSATION WITH FLUID POWER GIANT, JIM BRIZZOLARA

HydraForce was launched in 1985. In 1988, we started a joint venture over in England with our distributor there. A couple of years later, we bought them out, and so now that's a wholly owned subsidiary in Birmingham, England. In 2011, we started to build up our sales in Asia, and we opened up a facility in Changzhou, China, which is a couple hours outside Shanghai. This year, our distributor down in São Paulo, Brazil, had outgrown the two owners, and they asked us for some help. We decided that we would just buy them.

We've learned, primarily from our UK operation, that the UK has a very strong apprentice program. We've had apprentices in the building for many years in Birmingham. They come in the summer. We picked up on that in the U.S. about five/six years ago. This summer we had seventeen apprentices in here that we work with. One of the roadblocks you get in the U.S. is that manufacturing is considered a factory job. It tends to have a lower esteem than other operations. Finding people on the shop floor becomes difficult. If you need skilled engineering people, you can go to the local schools. We've always recruited in five or six local colleges. The last couple years, we started working very closely with local high schools here. Bringing the kids in, having them spend a day with us, seeing what we do, and trying to sell the high school kids on,"Hey a factory job isn't a bad job." It just may have a bad connotation, but it looks like it's a lot of fun. The people are skilled. The people are friendly. It's very diverse. It's not an assembly-line type operation.

Our big concern, when we went to China, was the turnover that's kind of a national standard there, I think. We started the operations up. Our first hire was a human resources manager. Our direction to her was, "You have to hire local people in the Changzhou area. They don't want people going home for the holidays, and not coming back. A year ago, in 2016, we had our fifth anniversary. We have a total of thirty two employees there, and eighteen of them were there for all five years. I was very proud of that. I was there, and I thanked them very much, and congratulated them. That's kind of a change there.

We can't let manufacturing, in this country, become complacent. They think that they're the strongest in the world, and there's no room for improvement. If you go back into the 1970s, you had General Motors, Ford, and Chrysler. They got that way. Look what happened when the Japanese came into this country. They almost put them out of business. The big three, really had become very complacent. They said, "Hey, we're here. We control it. We're as good as we need to be." The Japanese came in with some of the stuff. We have to be very very cautious that American Industry doesn't fall back into that trap that the big three fell into in the 1970s.

We are a global economy today. People are doing stuff all over the world. I think one of the problems is there probably aren't many people in the U.S. even interested anymore or that even understand, what tool and die making is. It's just kind of evaporated. If the people in the industry go away, they're not being replaced.

There are silos. They throw it over the wall. One of the things that we do here is when a new product is released, it's reviewed at the design stage. They go into a pilot run, where engineering is out on the manufacturing floor with the manufacturing engineers and the production people, putting the thing together, identifying the mistakes. We call it our production readiness process, and it's very very important.

Continuous improvement, to me, is a career-long activity. You can never get to a point and say, "Oh, we're good enough." That's a term I don't allow at HydraForce. People better not ever say to me, "Well it's good enough." Okay, you'll never get to that point where it's good enough. There's always, "It could be better."

A CONVERSATION WITH JOHN WINZELER
A GEAR HEAD FROM WAY BACK

One of the people who I met while in the industry and again when I became involved in the Technology & Manufacturing Association (TMA) was John Winzeler. John was a past TMA chairman (several contributors are past chairmen) and always dives into anything that will make a difference when it comes bettering our industry, and making more opportunities for our young people. The thing that John has done more than anyone else I have come across is bring an artistic flair to manufacturing. Walking through Winzeler Gear, John's family business, you will see gears photographed and assembled in the most unconventional way, in order to make a more beautiful presentation. When asked about this, John says that since most of his customers are in Europe, he tries to bring their mindset to his shop—and it shows. Boy does it show! I most undoubtedly feel that if more companies thought about how their plants and products are viewed, the perception of manufacturing would start changing—overnight!

Our business today is primarily automotive. The majority of our automotive customers, for the most part, are German and Austrian component manufacturers. Our business today is the products which end up in the interior of automobiles. The gears we make would make the windows go up and down, the seats go back and forth. The motorized accessories that work within a vehicle. We've got things that are in a steering wheel, behind an airbag, in the seat, in the door, in the roof, in the trunk.

We have engineers working to help our customers develop the products. Gears are a fairly unique universe, and carrying those products through prototype modeling, prototype tooling, and ultimately into production.

The Winzeler family all came from Switzerland. We've got precision in our blood. I think that what I see in the workforce, when we're dealing directly with Europe, is that we have interaction with more engineers that have a better appreciation for making than we do in the States. It's one of the reasons that we're in the product design business. It goes much deeper than just application engineering. It's doing complete calculations of gear systems, creating systems, in some cases, to fit within a black box. Few of our customers have those capabilities. The good news is that our customers have outsourced their brains in an awful lot of areas. They'll have to rely on a supply base to allow them to be able to develop sophisticated products, and execute them at high levels. Recently, I had a customer that asked us to collaborate with a peer competitor in Europe. They're doing in Europe what we do for them here. When I met with the co- presidents before Thanksgiving at our plant, here is what keeps them up at night: In the middle of Germany, they can't find a workforce.

To try and change the paradigm, the way people look at manufacturing, we need to do nothing more than what our European competitors do. I think our European competitors put a lot more energy into how their plants and the products look—that they don't just function. We're probably a lot less different than our European friends. Americans are not as interested, at least in manufacturing, in the aesthetics or some of the other things that we've done.

We have to make it easier for young people to get job shadowing and internships and meaningful experiences at an early enough age. Expose them to the right things—with the right people. I think it's getting the paradigm shift of guidance counselors, kids and parents. A career in a skilled trade or a manufacturing profession is not less than a college education; it's just different.

At the end of the day, if you're going to be a professional, you need both the hands and the mind.

It's just figuring out what you want to be before you grow up and get all this education. It's more difficult, once you're working, to get the education on the job or at night. I think the work that the TMA has done, with the education road map and trying to show the ways to become a professional, is great. I think the point is, how do you convince anyone that there can be a profession in anything? Whether that's making bread, or making molds, making machine parts, or whatever, it's a life long journey. It's not just something where you get an education and you're done. How do you create that curiosity and grit? That you can stand tall, and society says, "Wow, you're a machinist," or, "You're a mold maker." That's a wonderful, admirable thing to do, not, "Oh, you're a college dropout."

I think it's this huge marketing campaign, that manufacturing's alive and well and that there are good careers. It's no different than farming. We're not going to require the number of people we used to require, but the people required today are going to be much higher compensated. They have to operate with many skill sets. The technology has gotten way ahead of the people. We've never developed the appreciation in this country for working with your hands, unfortunately. I know the people that built our company originally, were from Germany and Switzerland. They were allowed in the country, and they had the skill sets. We were fortunate enough to be able to find them. There are many examples like that in the TMA. I think it's this big dollar marketing image campaign, backed up by an awful lot of support from the industry, to do everything that needs to be done. My thought would be that you need to get a spokesperson.

My thought is that CHAMPION Now!® is that organization that could do the marketing, and pull in a spokesperson who can make the impact needed in this country. Yes, it would require a great deal of money and unified effort to create such a marketing plan and to implement such. This is the vision of CHAMPION Now!®

A CONVERSATION WITH HELMUT ALBRECHT
MILITARY COMMUNITIES ARE A NATURAL FIT

In Chapter Eight, I mention meeting two people when speaking in Nassau County, Florida. The other was a person by the name of Helmut Albrecht. Helmut is very involved in STEM related education as the company he worked for (Science First) makes and sells experiential products for education applications. Helmut was born in Germany and has dual citizenship. He brings a unique understanding of the culture in Germany, its apprenticeship model, and the current culture in the United States with manufacturing careers.

Science First, is in science education. Most of our products are in the areas of physics, chemistry, biology and earth science. We also do some environmental science. The core of the product is experiential learning. We do not believe you can learn the forces of mechanics through software. You have to experience it. A lot of our STEM kits and our STEM products are going in this direction.

We compete with other counties or states in the United States for landing a larger manufacturer. One of the key questions is "What does your skilled labor pool look like?" When we pitch to different manufacturers, in the beginning, we just looked at Nassau County. We now look at a bigger picture. The taxable area has a fairly good skilled labor pool due to the fact that there are not a lot of apprenticeship programs. Next to San Diego, Jacksonville is a city in the United States where most of the Navy soldiers are going from their military life into their civil life. That creates a skilled labor pool, even when they don't call it an apprenticeship, they go through so much theory and practical training while they are on board a ship, that they become excellent skilled workers.

One of the companies near Mannheim (Germany) is John Deere. Every year John Deere has sixty slots for apprentices. They have 2,000 applicants. This just shows you how interested people are in Germany to become an apprentice at a well-known company,

The area I would like to bring up has to do with the parents and with the acceptance of blue-collar workers in Germany. About 60% of the youth are going through apprenticeship programs. Parents don't see it as something negative. If their son is going through an apprenticeship program over three years, they see the advantages. They see that they get a good education. They see that they have a great opportunity when it comes to their career and when it comes to their income. Acceptance, from a parent's point of view, and also, from a total culture point of view, is very very high.

For the parents, it would be fairly easy. If you look at what college tuition will be and how long it will take to pay back, not every parent wants to send their kids to college. With the parents, it's fairly easy to set up this formula.

In the United States, the internship model could be the right step in the right direction. That could lead, over time—over two or three or four generations—to something similar. If this internship model is growing, then the automotive industry, a Hyundai, Mercedes Benz, Toyota, Ford, Volkswagen (all of them have plants here in the United States) are coming together saying, "What should be our standard curriculum?"

You hear a lot of politicians talking about, "We have to bring jobs back to the United States." Now we are making almost a 360-degree circle in our conversation. Do we have the skilled workers, so that we can bring jobs back?

FAITH CAN PLAY A PART – WHAT WOULD JESUS BE? WWJB? A CONVERSATION WITH JAMES PRESTON

PASTOR OF KINGSWOOD UNITED METHODIST

My wife and I have attended Kingswood Methodist Church in Buffalo Grove since 1985. At a time that we were struggling with the leadership in the church along came a new pastor—James Preston. James brought a sense of humor and insight and modern day metaphors that made him relatable to everyone. I feel blessed to have met James and proud to call him my pastor—proud to call him my friend. I enjoy going to Chicago Bear games and have season tickets. Each season I try to take James to a Monday or Thursday night game. It's an opportunity for us to bond and enjoy a game together. James and I have held the American flag on the field together and gone down on the field with a field pass before kickoff.

As James and I were talking on the way to a Bear game. We were trying to make an analogy with Jesus and how he went against the culture of the times and gravitated toward those ignored and shunned from society. So many in today's culture look down on manufacturing careers and those who have found a home in them. Today if you cannot learn traditionally, you are pushed aside and suggested to go into something like welding, machining, etc. The fact is that these careers need people who like to work with their hands while also being good in math and science. Manufacturing needs the best and brightest. Just because someone cannot learn in the prototypical way, does not mean that they are not one of the best and the brightest.

The other biblical reference James made was how Jesus chose his disciples and how he taught them. The disciples "apprenticed" under Jesus. These spiritual leaders were chosen to learn the lessons by spending time with Jesus and witnessing his teachings as he carried them out.

The other analogy that I would make is that Jesus was told to be a carpenter, a craftsman when he was young. There are so many parallels in woodworking and metalworking. Fast forward over 2000+ years later. Would it be so far-fetched to see Jesus in today's world as a machinist? Two thought provoking concepts between Jesus' skilled trade and the manner in which he taught his disciples. This dialog between myself and James is inspiring to draw such parallels.

Today's culture is such that everyone is convinced that a four year formal education is the only path to success, when this is simply not the case—especially when you consider rising tuitions and more stringent admission requirements. Then there are also those who go that direction, and the market wants and needs something different than what their chosen degree offers.

"I am deeply committed to the concept of mentoring and apprenticing people into effective and productive vocational performance along with civil and faith commitment as well. Terry, I like your program and all that you have been doing in this venue for manufacturing vocations." - *James Preston*

CHRIS KOEPFER *Editor-in-Chief, Production Machining Columns Post: February 2018* [3]

Apprenticeships in U.S. manufacturing used to be pervasive throughout industry with scalable programs instituted in companies large and small. Where did they go, and why did they go away? In general, my take is two-fold: competition and cost. When I was coming up in the manufacturing industry working for a large machine tool builder, its apprenticeship program was second to none. Comprehensive, it provided a continuous chain of leaders, managers and, in many cases, executives who would lead the company ongoing. An employee completing the entire 8,000 hour program earned the equivalent of a four-year degree. Apprenticeships for many years acted, to borrow the baseball analogy, as the farm system for a company to cultivate future talent and place it where needed in the organization. The benefits were numerous, including candidates with very skilled levels, which could appropriately populate a variety of needs within the organization with the skill sets honed for the position available. A benefit of such an apprenticeship program was that as the needs of the company changed, the infrastructure for additional training was in place in house. Companies would exert control and influence over what was taught and try to match it with what was needed. In our economic structure, the costs for this were born by the company. For many years, domestic manufacturing existed in a sort of steady state of competitive balance. That was until the rest of the world found our markets and began to successfully penetrate those markets. Competitive pressures lead to cost-cutting, and apprenticeship programs were a relatively easy target. I was told, at the time, that since the effects of cost-cutting in-house training wouldn't be felt for years, the programs could be reconstituted. In fact, they were kicking the can down the road.

Moreover, apprenticeship programs are finding that they must start at a much more basic level than previous generations because the remedial needs of applicant's basic reading writing and math are so low. I'm also told that another dirty little secret about some of today's candidates is the use of drugs, but I will save that for another column.

One trend in apprenticeship-like programs producing results involves collaboration between local shops and local or regional technical schools. Often, it takes a phone call in a meeting to discuss the mutual benefits of working together. Once a relationship is established, the shops consult with the schools about their needs, and in turn, the schools have the knowledge to tailor their curriculum, to fulfill those needs. The collaboration works both ways, providing the area business with better equipped workers and the school with placement opportunities for its students.

Traditionally, many apprenticeship programs present more general topics that are germane to manufacturing. The thinking is that a base grounding in manufacturing can be a starting point to develop the specific skills needed. Taking that thinking a step forward can lead to co-op programs. I am aware of shops that have arrangements with local tech schools, and in some cases, high schools, which allow students studying specific topics to put that knowledge into context on the shop floor. The student is assigned a mentor in the shop and can be deployed to areas of the shop in need of extra help. In many cases, the student is paid while working as a co-op. Shop owners told me that this program is like a test drive for both parties. The shop can evaluate the student without the employment commitment, and the student can get the real world experience of the environment they will encounter.

The traditional formula for executing apprenticeships has changed. The idea of spreading the risk and cost among entities can contribute specifically to the goal of ensuring a secured line of supply for future workers. How it gets done depends on various factors, but the bottom line is it must get done.[3]

CHAMPION NOW!® – "APPRINTERNSHIP"

We have heard how important apprenticeships are in Europe. I submit that internships are vital to things changing in the U.S. I have been involved in many internship programs. To the point that I have thought long and carefully about the format of what I think a great internship program should be. The one program that is far and above better than ANY other program that I have worked with is RAND HAAS' MCIP program. I was an integral participant in his first program. Both apprenticeships and internships have dramatic impact on both cultures abroad and here on our shores. I have chosen to coin the term **apprinternship**. This allows the student to first test the waters at a young age with an internship. As interest and confidence in the metalworking craft grows, the student then transitions into a modified model of an apprenticeship. The modified model features a more American culturally acceptable model.

The next step for me is to brand them into two differently CHAMPION Now!® branded internships. CHAMPION MFG and CHAMPION ENG **apprinternships**.

CHAMPION MFG **apprinternship** has a pathway into the manufacturing world. This is for the young person who has no means or aspirations of going to college. These individuals would be working directly with companies and developing a relationship with the company with the immediate future of going into their workforce after high school graduation.

CHAMPION ENG **apprinternship** has a pathway into an engineering career, and as a result a four year college education. These individuals would be going to college and looking for three to four years of summer work as a means to augment their time in between the regular college semesters.

The **apprinternship** is a direct play on both words apprenticeship and internship. My point is that in order for the European apprenticeship to work effectively in the U.S., it must be changed. By implementing an internship WITH the apprenticeship, we can break through the deep rooted cultural barriers here in the U.S.

A CONVERSATION WITH LAZ LOPEZ
GET FIRST HAND EXPERIENCE BEFORE CHOOSING A PATH

Up until just recently, internships were something only available to college graduates or students towards the end of their college career. The truth is that they weren't very common. In the next decade, we're going to find that a workplace learning experience, such as an internship or apprenticeship, will be an expectation prior to high school. These experiences will be aligned to a pathway that a student is already engaged in through coursework, similar to some of the European models that you've referenced. Our most effective comprehensive high schools will deliver student opportunities to earn early college credit and industry credentials related to a field of interest, along with a capstone experience in a workplace that helps each student affirm post high school career plans. This approach moves the student's discovery process earlier, utilizing high school for exploring and finding a career of interest. This approach prepares students to graduate and enter a career path as opposed to a job. After two or three years fully involved in that particular career pathway, students leave high school not only affirmed but with a resume of experiences and networks of professionals to build upon.

Most people live and spend the rest of their lives in the same community. The quality of the community one resides in tomorrow is a direct result of the opportunities, education and skill set that you infuse in those young people today. We have a vested interest because these are going to be your future neighbors and workforce. All of us have a responsibility to ensure that every child, as they graduate, has the best chance to pursue the skills and credentials they need to be fully employable. Today's students will either be a contributing member of the community, or we're going to be helping to support them and their families.

A CONVERSATION WITH MARK HIBNER
LOOK AT WHAT'S POSSIBLE!

When you look at the opportunity those apprenticeships are giving to these students, I think if I remember all the details correctly, they got full insurance, they're getting 401K benefits. They get twelve paid holiday vacation days in the year. They get fourteen sick days paid. They're getting paid, starting out as a salary, at $26,000.00. Next year, twenty-seven or twenty-eight, and their third year, maybe $28,000.00. When they graduate with all their certifications, in that cohort of study, they're making $45-50,000 starting out. If you look in five years, these kids are making good money and the company has already given them about $180,000.00 (in wages, benefits and tuition). These kids are just starting down that pathway. This opportunity in manufacturing." One female student had a totally awesome opportunity. She did an internship with a company in the summer between her junior and senior year, and when she graduated, she got paid, starting out at, $19.00 an hour. They're willing to pay for her college education. It's changing kids' lives. I think that's what's really really cool.

During this chapter, we have heard from around the world what has worked and what needs to be tweaked. Going forward, we need visionary leaders to make a difference. Whether it be in inspirational cinematic visual movies, or an industry lead reaching out to schools and becoming engaged with as many educational advisory boards as possible, both are a means of making a difference in young people's lives. The facts remain that too many industry members sit on their hands when it comes to stepping outside of their plants and becoming involved in education. Granted, it will take the time and patience of the industry member, but it is well worth the time invested. On the other hand, there needs to be a means to effectively connect with the general public. Movies can inspire, and since Hollywood has not taken the lead, others need to step into a niche role and produce videos that tell a story that needs to be told!

VISIONARY LEADERS

"It's not about this year's balance sheet. Instead, it's about your workforce in the next five years."

–Jeremy Bout, Producer and Host of Edge Factor

I've met people in my career who, by example, have shown and proven that we need to give back to the next generation and community. These manufacturing superheroes have inspired me and, as a result, have given me the impetus to complete this project. Here is what we can learn from them.

I am embarrassed to say that, until early spring 2018, I did not know of a movie by the name of "Spare Parts." Thank goodness my neighbor, Mary Lou Smith, thought to tell me about this, otherwise I might have missed a great opportunity to acknowledge such a great example of many topics that I have written about in this book.

This movie featured some very notable and easily recognizable actors and actresses—George Lopez, Jamie Lee Curtis, Carlos Penavega and Marisa Tomei, to name a few. Released in 2015, this movie tells the true story of the Carl Hayden Community High School robotics team that rises to the most unexpected success. Inspired by a 2005 Wired magazine article by Joshua Davis, the storyline tells of a 2004 underwater robotic competition, and how a team of four undocumented immigrant students, with the mentoring and guidance of an engineer from industry, catapulted into fame with the most unlikely of wins against handsomely financed collegiate teams, with no shortage of support or resources.

In watching this inspirational movie, I thought to myself, "This is what the media should be promoting about manufacturing, STEM education and those who seem to have been forgotten in our culture." The true story is a testimony to what mentoring can do for young people who might not have had the opportunities otherwise. The movie is about a high school in Phoenix with a large percentage of Hispanic students with limited financial resources. The stage is presumably set for a very questionable future for those who come out of such an environment. What inspires me most in the movie is not only the success that the students achieved, but the other subject matters such as ethics, honesty, mentorship, frugality, diligence, perseverance, problem solving and accountability just to name a few. It is from these lessons that America's Greatest Champions are made!

CHAMPION NOW!® – EDGE FACTOR PARALLEL PATHS FOR INSPIRATION

At the MFG (Manufacturing For Growth) AMT 2011 meeting, I was able to meet Jeremy Bout. He officially introduced his first episode at the event—The Chilean Miners Rescue, a great human interest story of how a small machine shop in Pennsylvania was responsible for the amazing rescue of thirty three men trapped underneath the earth's surface in a mine. All of us watched glued to the TV, waiting for the miners to be brought up alive. Very few of us knew the behind the scenes story of Center Rock, Inc. who manufactured the device that drilled the tunnel that allowed the miners to escape alive! The storyline and film was spot on. Everyone loved it!

Later that night, I was able to introduce myself to Jeremy and explain some of my ideas, vision and passion for the industry. Jeremy explained how he had worked in Buffalo, NY in a machine shop and even programmed five axis CNC machining centers. He was fascinated with manufacturing, but a career choice led him to filmmaking. He went on to explain that both worlds would collide when he came up with the concept of Edge Factor. I enjoyed meeting and talking to Jeremy that night. Little did I know that there was more in store for us both.

After this event, I am not sure I even remember how Jeremy and I began communicating. I think, initially, it might have been emails, then a phone call or two back and forth. I began to ask for copies of his film to show at schools to young people. We continued phone conversations and came to believe that we were always on the same page with the mission, vision and message. I had told Jeremy, at MFG in 2011, that if we let his project fail that we would ALL be to blame, and the industry would get what we deserved. I had to figure out a way to support his project and films. This young man was way too talented, focused, and determined to slip through the cracks.

A CONVERSATION WITH JEREMY BOUT
MOVIE STORYTELLER EXTRAORDINAIRE

My early years were not marked by an understanding of what manufacturing meant. It wasn't until I was eighteen and in my early twenties that I was introduced to CNC and robotics advanced manufacturing technology. I fell in love with the technology but also what could be done with the technology. Manufacturing and innovation allows a community to completely reinvent itself. It allows a creative inventor to take an idea and bring it to market. It allows a company to form around the invention and then support families in a community.

And so, my goal with Edge Factor was to tell these stories in a way that really put the manufacturer as the hero. The rock stars of our stories would be the people who were truly looking to help people to actually change what was historically possible. The Chilean Miners' Rescue was a great example of that because that story was very critical with the mentality that up until that juncture, there were no case studies of anybody ever drilling down through 2,300 feet of rock. But when there was this really crazy story of these thirty-three men found alive in this tiny little room, it was manufacturers who were the true heroes of that because they reinvented technology on the fly over and over—problem solved nonstop—until they drilled successfully through, and the men were pulled out of the hole.

The true magic of that story was actually the machinists, the engineers, and all those on that manufacturing team who worked through the night and around the clock. So that's the essence of what we wanted to do in the early days with Edge Factor. But then something happened that took my vision and expanded what I originally saw, when I got a call, out of the blue. My phone number was available on my website at that juncture—my personal number—and I had an educator call me from the West Coast, and he said, "Look. I don't know who you are, but I just played the trailer for your film for my students today, and one of my students came up to me afterwards and said, 'Mr. Fitzpatrick, I'm gonna change my major, and I'm gonna focus on machining because that story really really redefined how I saw what I wanted to do for the future.'"

The caller went on to say that he wanted to help me in any way that he could, and it triggered something in me. As a young man coming out of high school, it was an accident, a summer job that turned into a manufacturing career. But there was something wrong—a consistent issue that was lurking in the shadows that I started to see all around me. While it was really really exciting as far as technology, if you went to a major trade show or you went into many companies and looked around, you realized, "Boy, there's a lot of gray heads in this facility. Where are the young people?"

This call marked a renewed interest in storytelling, but with a clear focus on workforce development. Our next big feature film was all about mountain biking, it was called Gnarly Metal. And I invited some educators to the set. I had about four or five leaders in education come out, and I filmed some short, little vignette videos at the same time as filming our feature, as we worked on the feature film. I didn't know what I was doing at that juncture, but I knew that I wanted to do something. At this point, Larissa Hofman joined the Edge Factor team and one by one we have been growing the team with people who want to dig into this issue and get their hands dirty with us.

I think that's really an important point. Edge Factor has been marked by the fact of what we were willing to do. We didn't just want to talk. We didn't just want to analyze this problem. We wanted to actually do stuff. We started to look at storytelling as a great thing because it would start with a good story, but that's not where we would end it. You don't watch the credits roll and walk away. No, you take that film, and you start to pull out teachable moments. Because when I was a young person doing my math, I always asked the question, "When in real life would I ever use this?" I think many people ask that question. And yet when you watch a story like Gnarly Metal, and you see these riders in the air, and you start to see them landing in these almost impossible moves—that they're pulling on in the forest and on the barge and the different places we filmed—you start to realize, "Boy, there is rotation. There is g-forces. There's gravity. There's physics. Oh man, there's so many teachable things that come alive."

I have been invited to Keynote events across North America and appear on different TV shows and radio shows. Over and over there are different needs that are being put in front of us to say, "Hey, you're a storyteller, but I've got this problem. Can you help us solve these problems?" We have really been navigating the scope of all this, from the spectrum of sitting on the board of Manufacturing Day®, meeting with many large companies to small companies and realizing that the depth of the workforce development challenge that we're facing.

All these silos of all these different groups of people from economic development to workforce development to this industry and that sector to this school and that private school and that homeschool—there are so many different people in a community that need to be part of this conversation about getting students aware of what their head and their hands could do... it is truly incredible.

To date, we have created over 60 feature and/or short films. We've worked with some phenomenal athletes. We have worked with the biggest names in the rock music industry. The biggest brands in the world have been featured through some of the stories related to them. At any given time, I have approximately eight films in the queue. All of them spawn many many teachable moments, and all of the content that we're creating from a film and storytelling perspective has an academic alignment to the standards that teachers have to teach to and is aligned to the needs for all the sectors that industry is concerned about.

Edge Factor has created a suite of tools for the workforce development conundrum that everybody's facing. And, of course, all the different stakeholders need to have different types of tools. So the goal over the next few years, certainly for us, is that we want to build on the fact that we're working in thousands of communities now. Edge Factor has worked in almost every state. Many people know us from the Rock MFG Day Kit that was used for Manufacturing Day®. Many people know us from the various initiatives and films that we've put out, but all of our future efforts are going into a local approach to a national problem. When we are working in a community, we bring the community stakeholders together. We equip them with the strategy and the tools that they need to go execute and gather metrics needed to review success. We're helping businesses get closer to their local schools. Ultimately in the next few years, when communities think about doing workforce development, we want Edge Factor and our stories comes to mind. We want them to use the tools, but we want them to also to tell their own stories.

They should look within ten to fifteen to thirty miles of where they're sitting at that moment. There are careers in their community. Students don't have to leave town to find great opportunities within their own community. Now Edge Factor is thinking and helping communities talk about regional solutions to the national problem.

My life is a tapestry of other people's wisdom, and my life is simply a reflection of the many people who have invested in me. From the beginning of Edge Factor, there have been many people who have spoken wisdom into my life. I look at my team of men and women who have been so willing to pour themselves into this project with me. As a team, we often look back to appreciate those who have spoken into our lives.

Edge Factor is now a leader in workforce development, we have actively worked in thousands of communities, but we see that our success in many ways, is simply the reflection of all the incredible people who have surrounded us. It is always great to look back so that you can face forward, equipped with lessons learned. It is so important to teach our kids to know their roots. You may not use the same technology as older people, but look back and you can learn from them; there's a lot to be learned. Our communities should not lose that. There is a generation of people who are exiting industry that are incredibly gifted and talented and we need to bring communities together, not break them apart. At our heart, this is what Edge Factor hopes to achieve in communities around the world.

As parents in teaching our children to be honest and engaged, we each have a roll to play,. As teachers, to be relevant and real-world, and as business leaders, we need to remember that we are to play a role and not wait to for people who will walk through the door. Much to do, but great work to be part of... Step up and link arms. There is room for everyone.

VISIONARY LEADERS IN THE WORLD OF MANUFACTURING

I have been in the business now for almost four decades. There are some people I have met who have definitely inspired me. I traveled extensively for all of these years meeting with hundreds of leaders in manufacturing primarily in the Midwest United States. I also traveled in a half of a dozen other states meeting people in all aspects of the manufacturing world. I have been very fortunate to meet some pretty incredible people. Each of those who I have mentioned shared a common trait—they were visionaries. They knew that the future of manufacturing depended on them. I was always intrigued as to how they would have time enough to make this a priority in their lives. The more I met with each and listened to their story, the more I wanted to dive in as well. Some of these people were customers, some were competitors, some I knew well, others had a reputation that preceded them. All of them had a profound effect on what I hope to be my contribution.

A CONVERSATION WITH BOB WEISHEIT
PRESIDENT OF ROBERT C. WEISHEIT COMPANY

Bob Weisheit and I met in 1985. I had moved down from Wisconsin after being in the business for about five years. Bob had been working for his dad as well. Bob and I hit it off early in our careers, and he has grown his father's business tremendously over the last 30+ years. Bob impressed me with his commitment to the local East Leyden High School. He would mentor the students who were in the machining program, and as a result, a tremendous number of his employees came through the East Leyden program. Bob was also very involved in the Technology & Manufacturing Association (TMA), and I even ended up following his lead onto the TMA Education Foundation, where I served for three years. Bob shared with me his thoughts on what motivates and inspires him to continue working with the East Leyden program.

I was very fortunate to grow up in a home with a father who was a tool and die maker. He was a successful, self-employed person. He was highly revered by his customers and family members. I believe that fifty years ago, when someone was a tool and die maker, they got more respect than they do today.

I got a drill press for a Christmas present when I was seven years old. I always got great satisfaction, at a very young age, building things. I built five boats as a hobby before I was twenty five years old at home in my garage. Each time I completed one people would comment, "Man, you built this?" It was as if you were a high school football star and you just won the state championship at Soldier's Field! Everybody's like, "Wow, this is awesome." So there was a lot of support around making things with your hands that somebody might not quite get if you do really well in school get straight A's or are an athletic standout. I mean, some people would know about it, but you can't walk around with your straight A report card and show it to everybody.

My Dad had two major customers. Neither of them ever checked his parts. He would just ship them in, and they would put them right into their products. I actually heard it myself from the customers. "You know, for the first twenty years we never checked your dad's parts when they came in. We were small, and we didn't have a bunch of inspectors. We would just take the parts and put them right in the product."

When you know this, you automatically have to set the bar higher for yourself because there's no double check. That was very fortunate instead of having worked in a place with an inspection department and final inspectors and the customers could check and recheck parts. I didn't set the bar, I just worked up to it. The customer and my dad set the bar. These are things we teach in our plant today, you, the machinist, not the inspector, are responsible for your quality. Today, we have 'dock to stock' customers, meaning our customers do not check our parts. It's scary for some people, but I've been living with that my whole life, started by my dad, and it continues today. In the Aerospace industry, our parts support life, death, and $20-$40 million aircraft. It's not like making washing machine parts and the water pump failed. No one is going to lose their life. We prefer to train people from scratch. Our motto is: If you are going to do something, take the time to do it right!

Where do we find people? We start working with them early (freshman and sophomores) in high school and prepare them for the work world. We want a clean slate, a clean sheet of paper, and we try and show them proper techniques from the ground up. The only negative of our method is its slow growth. We found that in doing that, we wound up retaining about 50% of the people we hired. When we hired people from the outside who were trained elsewhere, our retention rate was only 25%.

We found, early on, that we had a higher percentage of people who stay with the company if we got those people from scratch, or even a person who might've been out of high school for three, or four, or five years but didn't have a job with a future, and brought them in and trained them ourselves.

It's expensive because you're spending all this time training, and it might take two to four years before you've got a person really well situated to what he's doing and being productive. However, to lose three out of four guys who you hire who were not trained properly, there's cost there too.

We're not opposed to hiring people with experience. My motivation for my strong stance in education of people was self-serving—so I could supply more employees as our company grew over the years from one person to sixty five people. The other reason is to provide qualified workers for the machining industry that is key to the defense industry and the economic health of our country.

Get connected to your local high school, junior college, or four-year college. You have to be willing to walk out of work one day a month and go sit with other like-minded business people. I'm on two advisory councils and both have over ten companies that are at every meeting. These companies provide part-time jobs for students and full-time careers after graduation while also providing money for continuing education beyond high school.

The longest-term employee at my company has been with us for thirty-two years, and he graduated from East Leyden High School in 1986. He's still here and doing very well. He makes over $100,000 a year. There are twenty people here that came as a result of our high school interfaces. It does work, it does take time.

You go to a Bears game. The F-18s fly over and send the goosebumps up everybody's spine. When you know that plane doesn't fly without our parts, its very satisfying. intriguing. You get a lot of respect when you produce parts for a $40 million airplane whose only purpose is to defend and protect the U.S. and our allies. Whereas if you just say, "I work in a machine shop," then probably a lot of people think, "Well, when I went to high school thirty to forty years ago, all I remember is these guys had to wear these aprons and glasses because oil and chips were flying off the machines. It's certainly not very glamorous. There's were wood floors, and the machines were leftovers from World War II . . . "

Expose the parents! I remember bringing my wife to her first IMTS show four years ago, and she was awestruck for the whole time she was there. She couldn't believe the technology! She was married to me for all these years—had been in my plant, and had seen CNC machines since 1982 when we bought our first one. Even after thirty five years, when she went to that IMTS show, she was like,"It changes your whole perception."

If you want to get an outsider interested in the machining industry, bring them to an IMTS show or provide a plant tour. When possible, include the parents. They are a big influence on their children and will support the idea of a young person choosing this career if they know about the great satisfaction and economic success a career in this type of manufacturing can bring.

There is great satisfaction and financial reward to be had from a career in manufacturing. For example, you take a plain bar of metal and use your brain to figure out how, with the help of a $200,000-$600,000 computer-controlled machine, to craft that bar into the components of a jet plane that carries passengers safely across the continent. Or go into a missile or a rocket that will take out an incoming missile sent by an enemy nation (that if not intercepted could kill 500,000 Americans).

I think those who go into manufacturing careers because of supply and demand will make more money. There will be less people in it and, therefore, less competition and therefore their incomes will go up. I think the ability of any country to defend themselves is directly tied to their ability to manufacture their own products versus being at the mercy of or dependent on other countries for their defense equipment.

Manufacturing helps improve an economy—any manufacturing, whether you're in a war or you're not—because it's full-time employment and people are happier and can buy more thing. It's a circular effect. They have more money, then they buy cars, or they go on vacation, or they go out to eat, or whatever.

Go to college or go to work? My view, is you can go to college AND work in manufacturing if you really want to. I was able to do that. I worked about thirty hours a week and was a full-time student for four years of college—two in junior college and two in a four year school, Elmhurst College, where I earned my degree. No, I didn't go away to school. No, I didn't join a fraternity and have a party lifestyle and other things that some kids that go to college do. I worked and now own a very profitable business and a job I love.

To me, this is the ideal situation. I don't mean just manufacturing. If you want to be a carpenter, you can work with a carpenter crew during the day and go to night school. You're going to be so far ahead because you've got that combination of real world hands-on and the academic component to support it. This is the ideal path if you're willing to work that hard.

Only one or two of them (my employees) have a four-year college degree. The rest have associate college degrees, manufacturing engineering degrees. Some have only high school degrees. They all have very good jobs. That word needs to get out there, too. Part of the industry's fault is that we don't market well enough. People don't like to talk about wages with strangers or out in the open public. It is not uncommon for people at our company and in the machining industry to earn between $50,000 and $150,000 annually depending on their years in their career (10-40 years) and their individual skills and technical difficult of the parts they produce.

Somehow, that message needs to be part of the other things that we've already talked about; positive and safe work conditions, interesting work and well-paying jobs—it all goes together. People can find a place like that, they're around. They're in every city, people write articles about them. Request a tour, apply for a job. It very well may best the best thing you ever did for yourself.

These two visionaries have set the gold standard for all of us. We need more people like both of them to grab the reins and lead the way. In the final chapter, I give what I hope each sector needs for food for thought to become engaged to help of us in FINDING AMERICA'S GREATEST CHAMPION!

CHOOSE YOUR GAME PLAN: A PATH OF ACTION

"I tell young people – arrive early – leave late and TELL the TRUTH."

–Terry M. Iverson, Founder CHAMPION Now!®

This entire section is bulleted. These are the marching orders by group: for parents, youth, education, etc. It's a great summary piece that is a real call-to-action. Regardless of who you are or what you do, you should come away from reading this with a change that should be positive for yourself, your company, your child and your country. By now, you already know I started an organization named CHAMPION Now!®. We have defined the acronym as Change How American Manufacturing's Perceived In Our Nation. Despite that, this is intended for the USA, the A in the acronym can also stand for Advanced (instead of American) which means this can apply to any country in the world. Having covered the acronym itself, we have not spent a lot of time on the NOW! The Now is the call to action. Well, this chapter is the call to action.

EDUCATORS

Don't teach to the test! One of the biggest buzzkills in the world of education is the system that promotes teaching to the test. When this happens, we are destroying creativity and denying the most important skill that employers want and students so desperately need—problem solving skills. Young people today appear to be staying younger longer. Parents are over-protecting and trying to make things easier for their children. Meanwhile, teachers are teaching to the test, which is a double whammy when it comes to raising and educating confident, capable, and independent thinkers. We need our youth to lead. How is this going to happen when we are not allowing and or encouraging them to problem solve?

INVESTIGATE. See what is out there. In the manufacturing world, many educators are not privy to all the current great opportunities. Yesterday's careers are not necessarily hiring in today's marketplace. There are so many things that young people do not know. With technology changes, new job descriptions and positions are emerging every year. The market needs far less four year degreed workers and, instead, more skilled workers. Learning a skill is more in demand than the theoretical knowledge that we instruct our children and young people to focus on. Our culture needs to change. Our perceptions are antiquated. Educators need to talk to the industry partners and understand what they require, and then drill that down to the students. By investigating where the market has changed, and where the good paying jobs are, the educators can tell the students where their time and money should be spent. This should give them the biggest bang for their buck.

COMMUNICATE. Educators need to communicate with both their industry partners and with their students. By not listening and talking to both sides, we will get a vacuum. The end result will be that as the market changes, our educational model doesn't. Our culture also doesn't change. As a result, we end up having students spending a lot of time, energy and money to go into fields that either do not exist any longer, or there is such a saturation of applicants, that only the top single digit percentage of applicants are hired into good paying professions. The others may get interviews, but they are not able to differentiate themselves adequately to be employable. Students end up working basic service jobs and some not having any other choice but to live with their parents, not able to financially fend for themselves.

PROBLEM SOLVE. Young people today are very much lacking in this regard. I think that this is from a combination of parents trying to give and over-protect their children and the educational system not doing enough to educate individuals to be independent and able to analyze and come up with solutions for any given scenario. Unfortunately in the workplace, this is what employers want and so desperately need. In today's work environment, so many things change. When there is change, protocol goes out the window. Supervisors cannot be everywhere and need to trust that their subordinates who work for them can make an evaluation and choose a direction that is best for the company. By parents thinking they are doing the right thing and schools teaching to test so that they are evaluated only in a positive manner, many times the result is young people who have little or no problem solving skills, or leadership skills.

COUNSEL. This is one of the most under-appreciated positions in all of education. Young people are very poor in taking counsel for something as important as their secondary schooling and career path choices.

The people who they end up listening to are either Mom or Dad, and another influencer. Too many times, the counseling and direction they get is based on outdated data and misperceptions. Manufacturing is just one example as to a career path where there is far too little known. Despite hundreds of thousands of good paying jobs being available, most young people and their influencers have misperceptions of what manufacturing is and what jobs exist. Other outdated views include lack of technology and boring rote, task driven jobs.

PARENTS

My wife and I grew up together. We married at twenty one and twenty. One thing that we were able to both develop was the art of futuring. While this was not necessarily fair or enjoyable to our children, it made us better parents and better professionals. Futuring is when you think about the what ifs in any scenario in your day—in your life.

While my wife was thinking what if this happened what would I do? I was thinking what am I missing in any task? In any scenario, Kathy would be arranging for a plan B, while I was trying to constantly think how I could do something better, do something more completely.

Having said that, I could not be prouder of our three adult children. They are awesome! And, as you hear so many times, each and every one of them are different. This is so true. Each have their own style and unique skill set. The mainstay that runs through each of them is honesty. Each and every one of them are terrible liars. I am proud of many things, however, I am so very proud of that!

HONESTY, HARD WORK, ACCOUNTABILITY, INDEPENDENCE. These are just some of the building blocks our children need in life. The more of a foundation that we give our children, the better prepared they are for what life throws at them. And yes, life will throw all sorts of things to make it hard. Life is hard. We, as parents, need to give our young people the tools to slay whatever dragon is around tomorrow's corner, or behind a closed door.

EXPLORE Children need to know what their options are. Too many times we do not encourage our children to learn something such as a skill they may have been waiting to be developed. The more we expose our children to in terms of opportunities, the better. I am a big believer that knowing what you DO NOT want to do is the first step in learning what you WANT and ultimately WILL do in life. So many times, I hear young people saying, "Well that was a waste of time." This is not true. Learning what you need to develop, or do not enjoy, is valuable.

This is the first step to what you do well and what you enjoy. Internships are the perfect first step.

INSPIRE. One of the biggest needs a child has is to be encouraged. Most children are inherently not confident in trying new things. By inspiring our children, they can try whatever needed to fail or succeed. There is no guarantee that they will succeed at everything. Each and every young person has to be told it is ok to fail because without that, there is no concept of success and how to reach that pinnacle.

SUPPORT. Just inspiring without support is only half the battle. Our young people will fall, and we need to support them with resources and encouragement to get back up. Parents fill this role more than anyone in their lives. Many times, if there is a missing parent, some young people are so very fortunate enough to have the void filled by others like aunts, uncles, grandparents, coaches, mentors, etc.

EMPLOYERS - MARKET YOUR CAREERS AND YOUR CULTURE

I grew up in the world of manufacturing. I loved watching how things were made. Generally speaking, companies in the manufacturing sector have been acting as if there is something to be embarrassed about working in their companies. Somehow, they started believing that the U.S. was going to be a service related economy. Being proud and loud about their work world wasn't something that they thought was necessary or needed. If we let the media talk about how manufacturing is dead in the U.S., we will have allowed that to be a self-fulfilling prophecy.

"APPRINTERNSHIPS". In the future, maybe this is what they will become, but for now they are two separate programs.

INTERNSHIPS This is one area in this country that has taken off in terms of bringing new young people into the fold of many careers. Manufacturing has taken this almost to an art form level. Because manufacturers are terrible at marketing their careers ,and the culture's perception is so antiquated that the shortest distance between two paths is internships. The barrier that still gets in the way is that too many manufacturers assume that interns need to be eighteen or older in order to be accepted (safe) in a manufacturing environment.

APPRENTICESHIPS. The German Chamber of Commerce has done an awesome job of giving apprenticeships in this country new life. They are bridging the huge gap that exists between both the employers and the students. The American culture certainly is not what it is in Europe where the apprenticeship model is more accepted in their culture. While it is difficult to get on the apprenticeship train, do not give up so easily. There is a certain percentage of our youth that this is a perfect match for. The U.S. used to have an entire industry that embraced apprenticeships until failing economic circumstances forced many large corporations to give up their programs.

GIVE BACK. We need to get to the point that companies feel an obligation to give back and develop our young people into the leaders of tomorrow. Go outside your given field and help the company next door—someone you do not compete with. All too often, companies complain about the lack of an adequate workforce, but few have the courage and altruistic view to do something good just because it is the right thing to do.

PARTNERSHIPS. Too many employers are going the workforce development program solo. This has been something that has happened for decades. In order to survive, companies big and small in the U.S. have forced themselves to develop their own in-house training programs. While this is great, it is not as sustainable as when combined with some of the other models mentioned. Certainly, this will always exist in the U.S. culture, however how far ahead would we be if we could combine this with other concepts? Our employees would come into our interviews more capable and more confident. Our existing training programs will then pick up where they are and make them a polished skilled employee in much less time and at a much higher level.

MENTORS/COACHES

EXPECT EXCELLENCE. Pushing young people to be the best they can be is very important. Too many times, we reward children for showing up. While being involved is great, success is not merely measured by being present. Success is measured by results. Unfortunately, there has to be a winner. That is a simple fact. However, I would say there are no losers. I don't believe in that message. Losing is just the first component to eventually winning. How we win and teach young people to win and lose is important. When I coached, it wasn't about winning but, rather about continual improvement. As long as every player on the team got better each season, our team got better. But improvement doesn't just mean a won/loss record. It was also about learning life lessons and becoming more accountable, more mature and more independent.

CULTIVATE. So many times young people are looking for guidance. If we can cultivate our young people especially with the basic skills to be successful in life, their generation will be called the greatest generation. Maybe they will become the solution generation. I feel that young people have endless potential to become what they want to be. Environment is everything. I learned that at Bolles. We need to cultivate them for success. More are willing to listen and learn than those willing to mentor. Since I was so very fortunate to have mentors (besides my father who is an awesome mentor), I have felt the obligation to "pass this gift forward" and mentor others. More adults need to consider this. So many of our mature adults have talents, and we need to share these with the next generation, not just our own children, but beyond.

STUDENTS

Attitude is everything! I cannot tell you how many times I have been frustrated over attitude. I will take a great attitude over superior talent any day of the week. This one personality trait speaks volumes as to who you are. So many of my people in my company would run through a wall for me if asked. That is the ultimate compliment. The fact is, I feel the same way and try on a daily basis to support my coworkers.

EXPAND YOUR NETWORK. One of the reasons I wrote this book is to make a difference in the workforce, in the parenting world and for our youth. One of the biggest reasons I was able to write this book is my network. So many talented people have agreed to give me input and feedback. I consider many of these contributors to be some of the best in what they do in their given field. Each of them have a story. Each story speaks to someone. By allowing them to share their stories and their opinions, I am able to reach more people and make more of a difference in this world. Without my network, the outreach and range of this book would not be adequate.

WORK ETHIC. I have to admit that work ethic is one of the most undervalued traits. From what I have witnessed, the work ethic in the Midwest is one of the best in the U.S. I theorize it is because of the number of farm families, and it carries over into the culture of the area. In other parts of the country, I have seen that the work ethic is not quite as big and bold as the Midwest. Effort is everything. By trying our best, we learn how to be great.

PASSION. As a young person, I think that I was always trying to FIND what I was passionate about. I was a great student up until a time that I remember making a decision to be well-rounded. That meant not just being a good student and making good grades. I wanted to be a hard worker, a good athlete, a good friend and more social. By making this decision early in my career, I found I made choices and then lived with the consequences thereof. I remember my dad telling me, "Terry I do not care if you are out late having a good time. You can come in whatever time you want, BUT you better not let it show the next morning when you are at work. You give me your best and nothing less."

BE DRUG FREE . So many employers are shocked by the fact that applicants cannot pass a drug test. Even in my small company, in two of the most recent hiring opportunities, once I got to the final candidate, I told them that I would have a drug test requirement. We, of course, pay for the test. Both of these applicants said yes that they would take the test, accepted the information on where to take it, and YES neither of them showed up. Neither called back. This is a terrible problem in this country, and somehow we need to solve it. As a young person, make a commitment to yourself not to let this happen. You deserve better.

EXECUTE. What do they say? The path to failure is lined with good intentions. Don't let excuses define you. Let execution and results define you. Be tenacious to get the results you expect. This is not easy. Complacency and procrastination abound. I am a big advocate in environment developing and being responsible for bringing young people to a successful path in whatever they choose. If you are able to execute, then you will become the best you can be.

MENTOR YOUNGER PEOPLE, THEN YOU GET INTO YOUR PATH. The most effective way to mentor is to have young people mentor younger people. The response and end result is so much more effective. When someone older, who has been in the field for decades, mentors a young person, the continuity and relationship never gets past a certain point. The generational differences are so significant that this impedes effective progress in any mentoring program. Students mentoring students is a missed opportunity for everyone. Employers are going to be impressed with a new employee's leadership skills as evidenced by some sort of mentoring program. It is a win for the system and a win for you individually.

I recently watched a 1997 video of Steve Jobs from Apple talking about rebranding the Apple products. He spoke of many things, concepts and ideas, but there was one thing that he said that resonated greatly for me.

"People with passion can change the world for the better, and those crazy enough to think that they can, actually do. [1]

Many people that know me know that I am a passionate person and believe in making a difference. I ask you to do the same. Help me and CHAMPION Now!® make a difference for the young people of our country. Help with your time, your talents and your treasures - make a difference. Go to *www.championnow.org* and offer what you can. It could be by being a partner, by buying a book (or books), or spreading the word about exciting manufacturing careers.

Lastly, I tell all the young people the same thing when I meet them –"Arrive early—leave late and TELL the TRUTH" These eight words alone can take all of us a long way to success!

AFTERWORD

LEGACY VOICES

In Chapter One, I talked in detail of our family history and the beginnings of the manufacturing legacy that the Iverson family can stand proud to be a part of. The three brothers took the modest start that their father, Edward Iverson, gave them and grew the footprint to be quite vast and impressive. Below are comments from each of the three of them about their perceptions of change since they started in the business, their biggest accomplishments and what lessons and takeaways they have from their shared mentor and father.

ED IVERSON - Chucking Machine Products

When I started, all the customers I had—basically the vast majority—became friends. We trusted each other. We did almost everything on a handshake. Contracts or purchase orders didn't really have to be done. People's word was their bond. That's no longer true today, unfortunately. That's my opinion anyway. The other thing about the industry today is the major manufacturers; nobody wants to stick their neck out. They pass the buck so often that it just prolongs and makes things cost more. Things don't get done in a timely manner. There's a big difference. When they put out the Blackbird/the SR-71, which was maybe the fastest airplane we still know about, that plane went from blueprints to flying in twenty-seven months. They can't do it in ten years today. That's a big difference there, big big difference. It's not that people are dumber or anything. They just don't want to . . . When you make a decision, you got your neck out. It's much easier to pass things along and keep your head under the pillow.

The original Blackbird was designated the A-12 and made its first flight on April 30, 1962. The single-seat A-12 soon evolved into the larger SR-71, which added a second seat for a Reconnaissance Systems Officer and carried more fuel than the A-12. The SR-71's first flight was on December 22, 1964. The records set are many: The Blackbird was and remains the world's fastest and highest-flying manned aircraft. On its retirement flight from Los Angeles to Washington in 1990, to its final resting place in the Smithsonian Air & Space collection, the plane flew coast to coast in sixty-seven minutes. Most importantly, the aircraft delivered on its strategic responsibilities, providing the United States detailed, mission-critical reconnaissance for more than two decades. Only a select few know the true extent of the role the Blackbird's intelligence played in the Cold War. But its legacy as a game-changer will be admired for generations.[1]

I feel very strongly about what I said about the industry. Engineers, a good many of them, know the answers. They know they're on solid ground, but it's so much easier to pass the buck—pass it along. Then you don't have your neck out. If something happens, nobody's looking down on you. You're safe. Safe is what people want today. They don't want to be out there in front—where they should be. That was not the way our United States of America was built. People weren't afraid to make the decisions. A lot of them were tough, but they knew that to get things done, and done quickly and right, they had to do it.

I feel the biggest accomplishment with the company and the industry is that I virtually have never lost a customer. The main reason is that we do it right. If we do make a mistake for any reason, we own up and correct it. I treat all my people with honesty, and I don't BS them. Therefore I keep my people. I (Chucking Machine) started in '57, so what? Sixty-one years. I have a lot of nice equipment, but my biggest asset is my people. Without good people, you have nothing. I've treated my people and my customers the way I want to be treated. Today it's not always easy to do that with customers that aren't like they used to be, but you still have to do it.

The biggest lessons and takeaways I learned from Dad, I would say INTEGRITY. You don't hide anything under the rug. You own up to the good and the bad and the indifferent. Your word is your bond. You stay with it. I hope whoever reads what you do with this, I hope they get something out of it and learn from it. Ingenting å takke meg for (Norwegian — "Nothing to thank me for")

JERRY IVERSON - *Iverson & Company / Ternstrom & Company*

I started working for Iverson and Company part time in 1950 and full time in 1958. June of 1958. When I started into the workforce, the companies and the employees were much more loyal to each other. The companies took better care of their employees; the employees stayed with the companies much much longer and were more loyal to them.

Also, when I started out, the youth that came in from Europe had a lot of talent—a lot of machine tool talent. And even though they didn't have much money, they rose quickly in the workforce as far as an operator, a foreman, etc. Many of them went in to business and became quite successful. They were willing to take more chances than the present-day workforce, in my opinion. I put a lot of them in business through our leasing company, because they didn't have any money, but I knew they had talent. They wanted machinery to start in business, and I was willing to take a chance on them (lease them machines with virtually no credit). I had trust in their ability and in 95% of the cases, it worked. It was a good move for them and myself. I'm glad I did, I put a lot of young men into business who are very successful since.

My best accomplishment, by far, was profit sharing—to share my profits with the employees, to make sure their retirement was solidified and successful when they wanted to retire. By doing this, I got a tremendous amount of loyalty from my people, who started—some of them—in their teens and stayed with me until the day they retired. Of course without our employees there would not be any profits. That was, by far in my opinion, my biggest accomplishment. My other biggest accomplishment, as I said earlier, was putting a lot of young talented people into business—into the jobbing business through the leasing and having faith in their abilities without any credit ratings or very little cash to start their businesses, but they've all, 99% of them, become successful and built a good business.

The best lesson my dad taught me was to always be totally honest with your customers and totally honest with your employees. Don't fib, don't stretch the truth, be fair, be equitable, and be honest. Work hard. Set a good example for your people—which he always taught me. Be fair to your customer, and be fair to your employees. He also taught me that nobody is successful without good employees. You're only as good as your people, and if you think you can do it by yourself, you're foolish. You can't. You must have good people, and you've got to trust them. Otherwise, you'll never have a life of your own. You'll be working twenty-four hours a day. No successful businessman has ever succeeded without good people.

If you're going to go into machining, you've got to know programming. You've got to be able to do everything yourself when you start out. You can't go out and hire people to do it for you, because you won't have the money. In 99% of the cases, our customers all started out as individuals and then grew into employees and foremen. You don't start that way, so you have to have talent.

You have to have guts. You also have to be willing to put in the hours. You have to be willing to burn the midnight candle for a number of years before you get to a point where you can relax. There's no substitute for hard work. None. Every successful job shop that I know, every one of them, worked very very hard throughout their careers and were very talented when they started out, and then they were good teachers to their people and their sons and employees. But they had talent. They knew what they were talking about. They knew if they hired an employee whether that employee had talent or not because they themselves had talent. They knew the right questions and how to analyze an individual. So, that would be my recommendation to anybody starting out. Now you've got to have a little nest egg. If you try to do it without, you're probably going to fail. Today, you have to some backup money to fall back on. Nothing's going to come easy in the first two or three years. It's going to be very stressful. To start up, years ago, people used to go into it with very very little. Today, the cost of the equipment is so significant, the cost of employees is so much higher, you've got to have a nest egg to go into business. You've got to work for somebody, put money away on every paycheck until you feel you have enough to go out on your own.

JOHN IVERSON- Custom Products Corporation – KIST Industries

How is business is different today? You know, having started KIST Industries, it's so completely different. It's all been on the computer. The hiring is through the computer, finding the customers—virtually everything. You never have to leave your house. It's just completely different.

When I started Custom, I had to go out and hustle up every single customer individually, and that was face to face. It was just completely different. I think business had a lot more integrity, by far. We weren't raised to believe that business people are bad guys. I think it's sad because of Wall Street and the banks, I think kids grow up today thinking that business people are ruthless and cut-throat and all these terrible things. I never found that to be true at all in my career. In my memory, I don't really ever remember hurting another person or company, at least that I can remember or that I know of. Never did I intend to hurt anyone else. In my business, my competitors were constantly borrowing things in our quality department because we had one of the finest in the country. And I was happy to lend tools out, fixtures out and all kinds of things to all our competitors. In general, I thought it was a lot friendlier.

I idolized my father. For many years, I thought of him almost every day. And throughout life, I just tried to do whatever it took to make him proud. Even though he was dead all those years, virtually every decision I made was to make him proud of me as a son and to live up to his standards. He never said, John, that honor your word was everything. He didn't have to. It was just a given. His word was always good. He always followed through on his word, and I tried to do the same my whole life. I thought of my father every day for many many years.

My biggest accomplishment at Custom, I think right off, was that I bought a defunct company in 1971 that didn't have a single order, not a penny. When I sold, we were booked for $135 million; we had 800 employees and, a $22 million payroll. (Probably one of the most well-run companies in the country.) We were the largest machine shop in the country. There were two in Minnesota. They were second and third generation that were probably very close to our size, but they were fine companies. They, in almost all cases, had the same customers we did.

I was proud of the fact that I made parts for nine different IBM divisions. I was proud that I made parts for all the automotive companies, mostly in America, but also in Europe and some in Japan. Our biggest business was diesel, and we made the fussiest parts ever made in the world. We had tolerances of plus or minus a millionth of an inch. We could confirm that the thickness was in a millionth, the flatness, and all the true positions were within a few tenths of a thousandths. Those are tolerances that very few companies ever heard of. We made all the toughest parts that were made in the world. We made disc drive parts for every computer company in the country. I think one of the most accomplished things we did was that we had so many visitors. I actually retired because I felt like a diplomat. All I was doing was walking people around the various plants. We were disciples of Deming.

Dr. W. Edwards Deming Considered by many to be the master of continual improvement of quality, as well as their overall operation, Deming is best known for his pioneering work in Japan. Beginning in the summer of 1950, he taught top managers and engineers the methods for improving how they worked and learned together. He is often called the "father of the third wave of the industrial revolution." Playing a major role in the resurgence of the American automobile industry in the late 1980s, Dr. Deming consulted with corporations such as Ford, Toyota, Xerox,

Ricoh, Sony and Proctor & Gamble, whose businesses were revitalized after adopting his management methods. Deming was a visionary, whose belief in continual improvement led to a set of transformational theories and teachings that changed the way we think about quality, management, and leadership.[2]

We completely followed his philosophy in every respect. I laughed at companies who would come in all the time and say, "Oh yeah, we're gonna do that." And they're so full of bologna. It is a big deal to commit yourself to something like we did with those principles. We had to train all our suppliers. That's a big deal. It's very very expensive.

The company that bought my company had forty foundries/forty different companies. They used our quality systems in all forty companies, and they used our IT systems. 100% converted all forty companies to our computer systems. I'm proud of that.

The best years of my life were building Custom. I loved it. My best day of the week was going to work on Monday mornings. I really enjoyed it. Working with Susan (his daughter) . . . between you and I, when she'd walk in my office, it would just brighten up my whole day. She was just a joy to work with.

I never stopped being amazed at how dedicated the people were and how today, they really aren't. I mean, you walk into any store or ask any question of any employee, and they don't have a clue what they're doing. You can never have a happy relationship with your people unless you really train them. We learned that the hard way. We didn't have any turnover and, again, talking about accomplishments, after the first four or five months, we had almost zero turnover. We had sixty families where almost the whole family worked for me for basically their whole lives. They were so proud. One woman had a son who worked for me who stole something, and she quit. I called her, and I said, "You know, you're really a valued person in this company. We don't want to lose you." She was so dishonored by her son stealing that she couldn't stay. These are proud people. They're completely different people from what we have today, in my opinion.

A bunch of executives called me and said, "We are really in a jam. We've heard a lot of good things about your company and that you are somewhat of a gambler." They said, "We make bearings, wheel bearings for almost all the American automotive companies, and we can't make a part right now. All our machines are breaking down. We have machines on order, but they're a long way from delivery." So I bought, I think, forty chuckers (lathes to load and machine blanks).

We did all the automation, and it was a big deal. I said, "You know, I'm gonna do this, but I won't make money 'til the second year. If you don't give us a full two years, we're gonna lose our butts on this contract." A year went by. All during that year, they were gonna build statues of me. They just raved about us. Then they pulled the job. I said to them, "How can you do that? You knew what the deal was. I didn't get it in writing. I should have. I said, "You know, it's time for me to get out 'cause I've always been proud of the fact that manufacturing people are very very honest and reliable." The thing that is most noticeable to me today is that the MBA's today have been taught how to cut costs. It is also very noticeable to me that costs are everything and customers are second. The most noticeable thing is that companies don't put anything into training.

These three men have taught me almost everything that I need to know in business: dealing with customers, employees, vendors and how to be fair to all of the above. Few people in the world have the opportunity to be mentored by three so capable, talented and knowledgeable businessmen. I will be forever grateful for their leadership and advice. I know that their teachings have been passed onto the Iverson legacy of our three children and six grandchildren. An entire book could have been dedicated to the Iverson manufacturing legacy, but that could be for another day. I feel strongly the same as my two uncles and my father. We have all been so blessed to have great people work for us. When my Iverson & Company world almost ended in late 2009, it was the commitment and hard work from the people who had been with me my entire career that allowed me to push forward. Most everyone would have quit then and there. Probably even most people in my family would have. To this day, my grandfather's lessons have made me what I am, our children what they are, and our company's reputation what it is. The common bond all us have is Edward Alexander Iverson, my grandfather. We all owe everything to him. Thank you "Grandpa Senior!" You left such a legacy, and my hope is to make the most of your lessons. In my Uncle Ed's words above – I can imagine you saying...

Ingenting å takke meg for
(Nothing to thank me for)

NOTES

Preface

1. I am using the definition of a unicorn as found here on Divestopedia, https://www.divestopedia.com/definition/5114/unicorn.

Chapter 1: A Life in Manufacturing - The Iverson Path from Norway to U.S. Manufacturing

1. "National Signing Day' Isn't Just for Athletes: Ceremony Will Celebrate Students Entering Workforce," press release, Henrico County Public Schools, March 19, 2018, http://henricoschools.us/national-signing-day-isnt-just-for-athletes-ceremony-will-celebrate-students-entering-workforce/?highlight=National%20signing%20day. https://www.facebook.com/search/top/?q=Henrico%20County%20Public%20Schools%20National%20signing%20day

Chapter 2: Manufacturing: America's Gold Unicorn

1. I am using the definition of manufacturing as found here on Merriam-Webster, https://www.merriam-webster.com/dictionary/manufacturing
2. "Top 20 Facts About Manufacturing," website, National Association of Manufacturers, http://www.nam.org/Newsroom/Top-20-Facts-About-Manufacturing.
3. Bureau of Economic Analysis, fact No. 1
4. NAM calculations using IMPLAN, Manufacturers Alliance for Productivity and Innovation, fact No. 2
5. U.S. Census Bureau, Statistics of U.S. Businesses, fact No. 3
6. Internal Revenue Service, Statistics of Income, fact No. 4
7. Bureau of Labor Statistics, Fact No. 5
8. Bureau of Economic Analysis and Bureau of Labor Statistics, fact No. 6
9. Kaiser Family Foundation, fact No. 7
10. Bureau of Labor Statistics, fact No. 8
11. Deloitte and the Manufacturing Institute, fact No. 9
12. MAPI Foundation, using data from the Bureau of Economic Analysis, fact No.10
13. U.S. Commerce Department, fact No. 11

14. U.S. Commerce Department, fact No. 12

15. U.S. Commerce Department, fact No. 13

16. World Trade Organization, fact No. 14

17. Bureau of Economic Analysis, International Monetary Fund, fact No.15

18. Bureau of Economic Analysis, fact No. 16

19. Ibid, fact No. 17

20. Ibid, fact No. 18

21. U.S. Energy Information Administration, Annual Energy Outlook 2015, fact No. 19

22. Crain and Crain (2014) fact No. 20

23. "Top 20 Facts About Manufacturing," website, National Association of Manufacturers, http://www.nam.org/Newsroom/Top-20-Facts-About-Manufacturing. Deloitte and the Manufacturing Institute, fact No. 9

24. https://data.bls.gov/timeseries/LNS14000000 U.S. Department of Labor Percentage of Unemployment

25. Thomas Industrial Network, "Weekly Industry Crib Sheet: Boeing Forecasts Double Demand in 20 Years," June 18, 2013,https://www.engineering.com/Blogs/ tabid/3207/ArticleID/5872/Weekly-Industry-Crib-Sheet-Boeing-Forecasts-Double-Demand-in-20-Years.aspx.

26. Rick Romell, "Automation will prompt more employers to add jobs than to cut, Manpower says," Journal Sentinel, January 19, 2018, https://www.jsonline.com/ story/money/business/2018/01/19/automation-prompt-more-employers-add-jobs-than-cut-manpower-says/1048467001/.

27. "The Top 5 States for Manufacturing" is reprinted with permission from Manufacturing
Talk Radio, https://mfgtalkradio.com/top-5-states-manufacturing/
www.forbes.com/pictures/edgl45hfgm/no-13-atlanta-sandy-sp/#2e2742346491
chiefexecutive.net/the-top-10-states-for-manufacturing/
www.nam.org/Data-and-Reports/State-Manufacturing-Data/
www.bls.gov/eag/

28. Tony Schumacher wraps up his eighth NHRA season title in top fuel. by Jim Peltz NOV 15, 2014 | 8:15 p.m. http://www.latimes.com/sports/la-sp-nhra-finals-20141116-story.html

29. Associated Press, "Tony Schumacher breaks NHRA Top Fuel speed record," February 23, 2018, https://www.usatoday.com/story/sports/motor/nhra/2018/02/23/ schumacher-breaks-nhra-top-fuel-speed-record-force-returns/110766914/.

Chapter 3: Changing the Perception of Manufacturing: America's Greatest Business Opportunity

1. "Top 20 Facts About Manufacturing," website, National Association of Manufacturers, http://www.nam.org/Newsroom/Top-20-Facts-About-Manufacturing. Bureau of Economic Analysis and Bureau of Labor Statistics, fact No. 6

2. Manufacturing Day® website: About us www.mfgday.com/about-us

3. American Boilers Manufacturers Association website: www.abma.com/index.php?option...day...manufacturing-day...

4. The Manufacturing Institute. Updated April 2014, The U.S. Manufacturing Sector is the Eighth Largest Economy. International Monetary Fund and U.S. Bureau of Economic Analysis and MAPI.http://www.themanufacturinginstitute.org/Research/Facts-About-Manufacturing/Economy-and-Jobs/8th-Largest-Economy/8th-Largest-Economy.aspx

5. Jim Gribble, INFOGRAPHIC: "6 Myths About U.S. Manufacturing – Debunked!", October 20, 2017 https://www.marketing4manufacturers.com/6-myths-u-s-manufacturing-debunked/

6. Wikipedia

7. Bureau of Economic Analysis and Bureau of Labor Statistics

8. Bureau of Economic Analysis

9. Job Openings and Labor Turnover – June 2017 Release, Bureau of Labor Statistics

10. Summary Report for 49-9041.00-Industrial Machinery Mechanics

11. U.S. Commerce Department

12. U.S. Census Bureau, Statistics of U.S. Businesses

13. Kaiser Family Foundation

Chapter 4: Talent Void: In the Shop and the Classroom

1. 60 minutes interview with Tom Cook, December 20, 2015, https://www.youtube.com/watch?v=wdMxrovkpmU

Chapter 5: Key Influencers on Teenagers and Their Potential Futures in Manufacturing, Construction and Other Paths Unknown

1. Guidance and School Counseling-A Brief History of School Guidance Counseling in the United States, http://education.stateuniversity.com/pages/2023/Guidance-Counseling-School.html

2. MetLife foundation. "Poised to Lead: How School Counselors Can Drive College and Career Readiness" (December 2011, The Education Trust with Support from the MetLife Foundation)http://edtrust.org/wp-content/uploads/2013/10/Poised_To_Lead_0.pdf

3. Monte Whaley, "New law says students must be told about skilled labor, military careers
   Colorado public schools must inform high school students that not all post-secondary paths lead to college," The Denver Post, August 7, 2017, https://www.denverpost.com/2017/08/07/new-law-students-must-be-told-about-skilled-labor-military-careers/

4. I am using the definition of a maker as found here on https://en.wikipedia.org/wiki/Maker_culture

5. Jennifer Callaway and Yubing Shi, "Ignorance Isn't Bliss. The Impact of Opioids on Manufacturing:The intersection of the opioid crisis and manufacturing is poised to be a drag on U.S. competitiveness," February 22, 2018, https://mapifoundation.org/economic/2018/2/22/ignorance-isnt-bliss-the-impact-of-opioids-on-manufacturing

6. National Institute on Drug Abuse; National Institutes of Health; U.S. Department of Health and Human Services. Revised June 2015 https://www.samhsa.gov/data/sites/default/files/NSDUHresultsPDFWHTML2013/Web/NSDUHresults2013.htm#3.1.2

7. Drug-poisoning Deaths Involving Heroin: United States, 2000–2013 NCHS Data Brief No. 190, March 2015. https://www.cdc.gov/nchs/data/databriefs/db190.htm

Chapter 6: The Awesome Responsibility of Parenting All Without a Blueprint

1. Expecting Growth to Continue: The 2018 Construction Hiring and Business Outlook. Seventy-Five Percent of Construction Firms Plan to expand Headcount in 2018, Contractors are Optimistic about Strong Economy, Tax and Regulatory Cuts. January 3, 2018. https://www.agc.org/news/2018/01/03/seventy-five-percent-construction-firms-plan-expand-headcount-2018-contractors-are-0

Chapter 7: ROE: Return on Education Looking at Your Education Path as an Investment

1.   Charles J Sykes, "Fail U.: the False Promise of Higher Education," St. Martin's Press, New York, 2016, http://www.worldcat.org/title/fail-u-the-false-promise-of-higher-education/oclc/918994696&referer=brief_results

2.   Ibid.

3.   Student Debt Snapshot: A Current Picture of Student Loan Borrowing and Repayment In the United States.
      https://www.cometfi.com/student-loan-debt-statistics

4.   Drew DeSilver, "U.S. students' academic achievement still lags that of their peers in many other Countries," February 15, 2017, PEW Research Center, Washington, D.C., http://www.pewresearch.org/fact-tank/2017/02/15/u-s-students-internationally-math-science/

5.   Ibid.

6.   Thomas Snyder, "The Community College Career Track: How to Achieve the American Dream Without a Mountain of Debt," John Wiley & Sons, Inc., Hoboken, NJ, 2012

7.   Ibid.

8.   Ibid.

9.   Doug Belkin, "Why an Honors Student Wants to Skip College and Go to Trade School:As worries about student debt rise, states and businesses increasingly push faster, cheaper paths to the workplace; parents are stumped," Wall Street Journal, March 5, 2018,https://www.wsj.com/articles/college-or-trade-school-its-a-tough-call-for-many-teens-1520245800

10.   Drew DeSilver, "U.S. students' academic achievement still lags that of their peers in many other countries," February 15, 2017, http://www.pewresearch.org/fact-tank/2017/02/15/u-s-students-internationally-math-science/

Chapter 8: Leading the Way: Florida and Wisconsin

1.   http://www.nam.org/Data-and-Reports/State-Manufacturing-Data/State-Manufacturing-Data/April-2017/Manufacturing-Facts---Wisconsin/

2.   Florida Manufacturing Facts. U.S. Bureau of Economic Analysis and the U.S. Census Bureau.http://www.nam.org/Data-and-Reports/State-Manufacturing-Data/State-Manufacturing-Data/April-2017/Manufacturing-Facts---Florida/

3.   Https://www.census.gov/popclock/ https://www.census.gov/search-results.html?q=Florida+population&page=1&stateG eo=none&searchtype=web&cssp=SERP

Chapter 9: Find Your Passion, Design Your Life

1.   Jeffrey J. Selingo, "College (un) Bound the Future of Higher Education and What it Means for Students," New Harvest Houghton Mifflin & Harcourt, 2013.

Chapter 10: The Solution Generation's Bedrock

1.   D207 College and Career Readiness https://vimeo.com/110499612 ;
      Dr. Audrey Haugen Principal Maine West Principal,
      Dr. Michael Pressler Principal Maine East,
      Mr. Shawn Messmer Principal Maine South
2.   Steve Minter, "Will Millennials Change Manufacturing?" Industry Week Magazine, December 28, 2017, https://www.industryweek.com/talent/will-millennials-change-manufacturing

Chapter 11: Women in Manufacturing - Changing the Lens of What Manufacturing is

1.   Monte Whaley, "Colorado Families Struggle to Pay Skyrocketing Back-to-school Costs," The Denver Post, August 14, 2017, https://www.denverpost.com/2017/08/14/back-to-school-costs-colorado/.
2.   Mae Jemison, "First Woman of Color in Space, Talks STEM Gaps and Science Fiction," The Seattle Times, originally published June 21, 2017 at 4:00 am, updated June 21, 2017 at 11:00 am, interviewed at the Sheraton Hotel,https://www. seattletimes.com/education-lab/qa-mae-jemison-first-woman-of-color-in-space-talks-stem-gaps-and-science-fiction/

3.   Millie Dresselhaus, The University of Chicago. Accolades. Presidential Medal of Freedom, https://www.uchicago.edu/about/accolades/23/

Chapter 12: The Culture of Manufacturing from a Global Perspective

1.   U.S. Census World Population Clock. https://www.census.gov/popclock/
2.   Chris Koepfer, Production Machining Column, "Make an Impression," October 1,1997, https://www.mmsonline.com/columns/make-an-impression
3.   Chris Koepfer, Production Machining Column. Customized Apprenticeships. The reality of skill levels for today's potential apprenticeship candidate is hitting home, January 18, 2018, https://www.productionmachining.com/columns/customized-apprenticeships

Chapter 14: Choose Your Game Plan: A Path of Action

1.   Steve Jobs, Apple. https://youtube.com/watch?v=dR-ZT8mhfJ4

Afterword

1.   Creating the Blackbird.
     https://www.lockheedmartin.com/en-us/news/features/history/blackbird.html
     SOURCES AND ADDITIONAL READING
     Boyne, Beyond the Horizons
     Johnson and Smith, Kelly—More than My Share of it All
     Francillon, Lockheed Aircraft since 1913
     Rich and Janos, Skunk Works
     Lovick, Radar Man—A Personal History of Stealth
     Jacobsen, Area 51--An Uncensored History of America's Top Secret Military Base
2.   Deming the Man. The W. Edwards Deming Institute.
     https://Deming.org/Deming/Deming-the-man

ACKNOWLEDGMENTS

Inspiration comes in many forms. Sometimes when you least expect it. I have been inspired by so many people, and I hope to inspire some as well. The entire "Pay it Forward" concept is awesome. Movies can be very moving and inspirational. In 2000, a movie hit the screens and started a movement that is still alive and well today. Pay it Forward presented the concept of being the recipient of a good deed, and paying it forward to someone else. This concept I think pertains to mentoring and helping young people in today's complex, fast moving world.

I have been inspired by different people in mentorship, including:

- *My dad Jerry Iverson*
- *My uncles Ed and John Iverson*
- *My father-in-law Earl Thomas*
- *My teachers Frank (Boo Boo) Anderson and Buddy Ward*
- *My coaches Jim Kane, Dave Kenson, and Bob Swindell*
- *My elders at church Dick Stone and Dwight Hall*
- *My co-worker John Comparini*
- *My customers and friends Ben Shor and Kevin Sinnett*

Unfortunately I am sure I have inadvertently left people out, and for that I am truly sorry. ...

Thank you to all of my employees who carry the Iverson & Company flag each and every day:

Dale, Mike, Jim, John, Chris, Roger, Maria, Carolyn, Jason & Steve. Without you, we cannot succeed and I would not have been able to write this book.

Thank you to my creative team:

- *Cover and layout design: Juan Pablo Ruiz*
- *Developmental editor: Michele Kelly*
- *Publishing: Jackie Camacho-Ruiz/Fig Factor Media*
- *Proofreading and Copyediting: Teresa Bondavalli, Carol Lezak*

ONE LAST THOUGHT

For years, I would take my children to sporting events (Bears games), and many times I would give a dollar to someone less fortunate. Whether the people coming up to us and their stories were true or not made no difference to me. Whether they really did need the $19 for bus fare, or whether they instead used it for less than honorable purposes, I cannot tell you. It really doesn't matter because there will be those times that one will use the money for food, or for bus fare, or for something truly needed.

Even more than money, we can pay knowledge forward, and with knowledge comes opportunity. There were many who I talked about in this book who have given back to technical education and helping young people. So I would like to expand the concept to instead of giving money, giving knowledge. I didn't write this book to make money—hardly. What I did do though, is write this book to make a difference. To give back to this country. To invest in our next generation.

I challenge anyone and everyone in our industry, in any manufacturing to offer your time, talent, and treasures to CHAMPION Now!® Go on the website (www.championnow.org) and buy a book, or donate to CHAMPION Now!® We are one recession away from losing the momentum, and then it will be all too late to make the gains needed to increase our skilled workforce.

And, lastly, if you are given a book, use it, then give it to someone else who has crossed paths with you in your life and also needs it. That is my hope to make a difference in lives, and as a result start changing the culture in our country one person, one book, one life at a time.

ABOUT THE AUTHOR

Terry M. Iverson is President/CEO of Iverson & Company, a machine tool distributorship and rebuilder in Des Plaines, Illinois. Terry has been calling on machine shops and manufacturing companies since 1980 primarily in the states of Wisconsin, Illinois and Indiana. His family has a few family businesses in manufacturing-related areas. His brother, Erik, runs their company that manufacturers dial indicators (mechanical and electronic). Both companies are over 86 years old. Terry's two uncles both have/had sub-contract machine shops, while his father ran the company during his tenure, after his grandfather founded it.

Terry has spent thousands of hours canvassing the country, talking to schools and visiting with national and local state politicians about manufacturing in America. He has served on many local high school, community and technical college advisory boards. He has served on the TMA Education Foundation in Illinois, the CTE Education Foundation in Washington DC, and chaired the National Visiting Committee for FLATE (Florida Advanced Technological Education Center) in Florida. Terry has also testified to the Small Business Committee of the House of Representatives, in Washington DC.

Terry's wife Kathy and he have been married for 38 years after meeting each other in their junior year in high school. Both grew up in Jacksonville, Florida.

He attended mechanical engineering school at both University of Wisconsin Madison and Marquette University in Wisconsin. Terry and Kathy have three adult children each married to the love of their life. They have been blessed with three grandsons and three granddaughters. Terry coached travel soccer for 23 years, which each of his three children participated in growing up. He also played over 40 soccer until 2007, and was also a founding member of Grove United Soccer Association in Buffalo Grove, IL.

Terry founded CHAMPION Now!® with the vision of changing perceptions of manufacturing careers, in hopes of assisting to solve the skills gap crisis in this country. CHAMPION Now!® is an acronym that stands for Change How American/Advanced Manufacturing's Perceived In Our Nation. The "Now" references our collective call to action. Learn more about CHAMPION Now!®: www.championnow.org

19389497R00168

Made in the USA
Lexington, KY
28 November 2018